Lecture Notes in Mathematics

A collection of informal reports and seminars
Edited by A. Dold, Heidelberg and B. Eckmann, Zürich

261

Alain Guichardet

Faculté des Sciences de Poitiers, France

Symmetric Hilbert Spaces and Related Topics

Infinitely Divisible Positive Definite Functions
Continuous Products and Tensor Products
Gaussian and Poissonian Stochastic Processes

Springer-Verlag
Berlin · Heidelberg · New York 1972

AMS Subject Classifications (1970): 46 C 10, 46 M 05, 60 G 15, 81 A 20

ISBN 3-540-05803-6 Springer-Verlag Berlin · Heidelberg · New York
ISBN 0-387-05803-6 Springer-Verlag New York · Heidelberg · Berlin

Offsetdruck: Julius Beltz, Hemsbach/Bergstr.

Table of Content

INTRODUCTION

As was recognized by J.M.Cook in 1953 the notion of symmetric (or exponential) Hilbert space, which is quite similar to that of the symmetric algebra of a vector space, is fundamental to a construction made in 1932 by the Russian physicist V.A.Fock in order to provide a more rigorous foundation for Quantum Field Theory. This construction, which is now called the Fock representation of the canonical commutation relations, is presented in § 2.2 and Appendix F. More recently the Japanese physicist H.Araki supplied a new impetus to the theory of symmetric Hilbert spaces, showing that it is related to infinitely divisible positive definite functions on groups and provides a new method for constructing all these functions on a given group. These ideas were further developed by K.R. Parthasarathy, K.Schmidt and R.F.Streater ; we present them in chapters 3 and 4. Chapter 5 is devoted to another deep result of Araki concerning what we call Boolean algebras of tensor decompositions of a Hilbert space, which shows that symmetric Hilbert spaces play to some extent a universal role in the theory of continuous tensor products of Hilbert spaces. This result is applied in chapt.6 to the determination of the so called factorizable positive definite functions on some groups which can be considered as continuous products of groups.

From another direction J.Neveu and independently G.Kallianpur established recently a link between symmetric Hilbert spaces and Gaussian stochastic processes, giving a new interpretation of the works of Wiener, Ito, Cameron and Martin about the so called Wiener stochastic process. We present these ideas in chapt.7 where we have felt it useful to make a systematic exposition of the theory of Gaussian measures on topological vector spaces. There is also a relation between symmetric Hilbert spaces and Poissonian processes, but a less direct one, by means of the more elementary notion of symmetric set of a set which we introduce in chapt.1. Finally in chapt.8 we discuss various possible notions of continuous products which are potentially useful but which are not as yet quite precise, and we try to determine the directions in which one can search for precise definitions.

After writing these notes we learned that K.R. Parthasarathy and K. Schmidt gave at the same time in London a lecture on a similar subject entitled "Positive definite kernels, continuous tensor products and central limit theorems of Probability Theory".

Chapter 1. THE SYMMETRIC MEASURE SPACE OF A MEASURE SPACE

In this chapter we introduce several elementary notions which are rather similar to that of symmetric Hilbert space.

§ 1.1. The symmetric set of a set

Let X be a set ; for each $n = 1,2,\ldots$ we denote by S^nX the quotient set of X^n by the symmetric group $\mathfrak{S}_n$ acting on X^n by permutations ; S^oX is by definition reduced to a single point ω .

Definition 1.1. The set-theoretical sum (or disjoint union) $\sum\limits_{n=o} S^nX$ will be called symmetric set of the set X and denoted by SX.

For every n-uple $(x_1,\ldots x_n)$ in X^n we denote by $cl(x_1,\ldots x_n)$ the class of $(x_1,\ldots x_n)$ in S^nX ; then we define a composition law in SX by

$$cl(x_1,\ldots x_n) + cl(y_1,\ldots y_p) = cl(x_1,\ldots x_n,y_1,\ldots y_p)$$

$$\omega + cl(x_1,\ldots x_n) = cl(x_1,\ldots x_n) + \omega = cl(x_1,\ldots x_n) ;$$

SX becomes a commutative monoid with unit element ω, the so called free commutative monoid with unit generated by X. Identifying S^1X with X we can also write

$$cl(x_1,\ldots x_n) = x_1 + \ldots + x_n .$$

The canonical injection $X \longrightarrow SX$ will sometimes be denoted by φ_X. We can also describe the monoid SX in the following way : let $\underline{N}^{(X)}$ be the set of all mappings $X \longrightarrow \underline{N}$ such that $N(x) = 0$ almost everywhere (i.e. except for a finite number of x) ; $\underline{N}^{(X)}$ is a commutative monoid for pointwise addition ; there is an isomorphism of SX onto $\underline{N}^{(X)}$ which carries each element $x_1 + \ldots + x_n$ into the function N where N(x) is the number of indices $j = 1,\ldots n$ such that $x_j = x$. We shall refer to the first and the second description respectively as the S-<u>picture</u> and the N-<u>picture</u>.

§ 1.2. The functor S on the category of sets

The monoid SX has the following universal property : if G is a CMU (commutative monoid with unit) every mapping $u : X \longrightarrow G$ extends to a unique morphism $\mathring{u} : SX \longrightarrow G$ defined by

$$\left.\begin{array}{l} \mathring{u}(x_1 + \ldots + x_n) = u(x_1) + \ldots + u(x_n) \\ \mathring{u}(\omega) = 0 \end{array}\right\} \qquad (1.1)$$

Therefore S can be considered as a functor of the category $\mathcal{S}$ of sets into the category $\mathcal{CMU}$ of CMU : if we have two sets X and Y and a mapping $f : X \longrightarrow Y$ there exists a unique morphism $Sf : SX \longrightarrow SY$ such that the diagramm

$$\begin{array}{ccc} X & \xrightarrow{\;f\;} & Y \\ {\scriptstyle \varphi_X}\downarrow & & \downarrow{\scriptstyle \varphi_Y} \\ SX & \xrightarrow[\;Sf\;]{} & SY \end{array}$$

is commutative ; Sf is defined by $Sf(x_1 + \ldots + x_n) = f(x_1) + \ldots + f(x_n)$. In the N-picture Sf is given by

$$(Sf(N))(y) = \Sigma\, N(x)$$

where x runs over $f^{-1}(\{y\})$.

The functor S is coadjoint to the functor $F : \mathcal{CMU} \longrightarrow \mathcal{E}$ which carries every CMU into the underlying set (for the definition and properties of coadjoint functors see e.g. [29] ch.II, § 12) ; in fact we have for each set X and each CMU G a bijective correspondence

$$\mathrm{Hom}_{\mathcal{S}}(X, FG) \longleftrightarrow \mathrm{Hom}_{\mathcal{CMU}}(SX, G)$$

with the required properties. As a coadjoint functor S transforms sums into sums ; but the categorical sum of a family of CMU $(G_i)_{i\in I}$ is nothing but the restricted product

$$\prod_{i\in I}{}' G_i = \{ (g_i) \in \prod G_i : g_i = 0 \text{ a.e. } \}$$

More precisely let $X = \sum_{i\in I} X_i$ be the sum of a family of sets X_i ; there is an isomorphism $SX \longrightarrow \prod{}' SX_i$ carrying every element $\sum_{i,j} x_{i,j}$ with $x_{i,j} \in X_i \ \forall j$ into the family $(\sum_j x_{i,j})_{i\in I}$. In the N-picture this isomorphism carries every function N into the family $(N \mid X_i)_{i\in I}$ where $N \mid X_i$ is the restriction of N to X_i.

The set $\underline{C}$ of complex numbers being both an additive and multiplicative CMU, every mapping f of a set X into $\underline{C}$ can be extended in two different ways into mappings $SX \longrightarrow \underline{C}$; we denote them respectively by $\overset{\circ}{f}$ and $\overset{\circ\circ}{f}$; more precisely

$$\left.\begin{aligned} \overset{\circ}{f}(x_1 + \dots + x_n) &= f(x_1) + \dots + f(x_n) \\ \overset{\circ}{f}(\omega) &= 0 \end{aligned}\right\} \qquad (1.2)$$

$$\left.\begin{aligned} \overset{\circ\circ}{f}(x_1 + \dots + x_n) &= f(x_1) \times \dots \times f(x_n) \\ \overset{\circ\circ}{f}(\omega) &= 1 \ . \end{aligned}\right\} \qquad (1.3)$$

In the N-picture we can write

$$\overset{\circ}{f}(N) = \sum_{x\in X} N(x)\, f(x) \qquad (1.4)$$

$$\overset{\circ\circ}{f}(N) = \prod_{x\in X} f(x)^{N(x)} \qquad (1.5)$$

in particular we have $\overset{\circ}{0} = 0$, $\overset{\circ\circ}{0} = \delta_\omega$, $\overset{\circ\circ}{1} = 1$. Now suppose that X is a sum $\sum_{i\in I} X_i$; set $f_i = f \mid X_i$ and $N_i = N \mid X_i$; then

$$\overset{\circ\circ}{f}(N) = \prod_{x\in X} f(x)^{N(x)} = \prod_{i\in I} \prod_{x\in X_i} f_i(x)^{N_i(x)}$$

$$= \prod_{i\in I} \overset{\circ\circ}{f_i}(N_i)$$

so that we can identify $\overset{\circ\circ}{f}$ with the function $\bigotimes_{i\in I} \overset{\circ\circ}{f_i}$.

§ 1.3. The symmetric Borel space of a Borel space

In this paragraph we suppose that X is a standard Borel space (concerning Borel spaces see [9] App.B or [32]) ; we endow X^n with the product Borel structure, S^nX with the quotient of the latter, and finally SX with the sum Borel structure of all S^nX ; X being standard there exists a total order relation $\leq$ on X which is Borel in the sense that for every n the set X'^n of all n-uples $(x_1,\ldots x_n)$ satisfying $x_1 \leq x_2 \leq \ldots . \leq x_n$ is a Borel subset of X^n ; then S^nX is Borel isomorphic to X'^n, hence standard, and SX is standard too ; moreover it is easy to verify that SX is a Borel monoid (see App.D.1).

In short we have associated with every standard Borel space X a standard Borel monoid SX ; moreover the canonical injection φ_X : X $\longrightarrow$ SX is Borel. There are properties quite similar to those of § 1.2 : if G is a Borel CMU every Borel mapping u of X into G extends to a unique Borel morphism $\overset{\circ}{u}$ of SX into G defined by (1.1) ;

S is a functor of the category of all standard Borel spaces to the category of all standard Borel CMU ; it is coadjoint to some functor, therefore it transforms sums into sums, in other words if X is the Borel sum of a countable family X_i, SX can be identified with the Borel restricted product $\prod' SX_i$ (see App.D.1).

§ 1.4. The symmetric measure space of a measure space

In this paragraph we consider a standard Borel space X and a (finite positive Borel) measure μ on X. For every n we denote by μ_n the image of $\mu^{\otimes n}/n!$ under the canonical mapping $P_n : X^n \longrightarrow S^n X$; μ_n is a measure on S^nX having total mass $\|\mu_n\| = \|\mu\|^n /n!$.

Definition 1.2. We denote by $S\mu$ the measure on SX, the restriction of which to S^nX is equal to μ_n for every $n \geq 1$; on the set $S^oX = \{\omega\}$ we define $S\mu$ by $S\mu(\{\omega\}) = 1$. We thus have

$$S\mu = \bigoplus_{n=o}^{\infty} \mu_n$$

$$\|S\mu\| = e^{\|\mu\|}$$

$$< S\mu, f > = \sum_{n=o}^{\infty} (n!)^{-1} < \mu^{\otimes n}, f \circ P_n >$$

where f is an arbitrary bounded Borel function on SX. We also have $n!\,\mu_n = \mu^{*n}$ = the n-th convolution power of μ considered as a measure on the monoid SX, and $S\mu = e^{*\mu}$ = the convolution exponential of μ. If $\mu = 0$, $S\mu = \delta_\omega$; moreover $S(\mu+\nu) = S\mu * S\nu$. If X is reduced to a single point a and $\mu(\{a\}) = 1$, SX can be identified with $\underline{N}$ and $S\mu$ with the Poisson measure $\sum_{n=o}^{\infty} (n!)^{-1} \delta_n$.

Properties of $S\mu$.

(i) Suppose that μ is non-atomic ; let $\leq$ be a Borel total order relation on X and define X'^n as in § 1.3 ; then if we identify S^nX with X'^n, μ_n <u>is identified with</u> $\mu^{\otimes n} | X'^n$. In fact let Y be the subset of all n-uples $(x_1, \dots x_n)$ in X^n such that $x_i \neq x_j$ for $i \neq j$; $X^n - Y$ is $\mu^{\otimes n}$- negligible since μ is non-atomic ; if A is a symmetric Borel subset of X^n, $A \cap Y$ is the disjoint union of the subsets $\sigma(A \cap Y \cap X'^n)$ where σ runs over $\mathfrak{S}_n$; then

$$\mu^{\otimes n}(A) = \mu^{\otimes n}(A \cap Y) = n!\ \mu^{\otimes n}(A \cap Y \cap X'^n)$$
$$= n!\ \mu^{\otimes n}(A \cap X'^n)$$

and on the other hand

$$\mu^{\otimes n}(A) = n!\ \mu_n(P_n(A)).$$

(ii) If u is a Borel mapping $X \longrightarrow Y$ we have $S(u(\mu)) = Su(S\mu)$ where Su is the mapping $SX \longrightarrow SY$ associated with u as shown in § 1.2, $u(\mu)$ and $Su(S\mu)$ are the images of μ and $S\mu$ respectively under u and Su (easy verification).

(iii) For every $f \in L^1(X, \mu)$, $\overset{\infty}{f}$ belongs to $L^1(SX, S\mu)$ and

$$< S\mu , \overset{\infty}{f} > = \exp(< \mu , f >). \tag{1.6}$$

In fact for every n, f is μ_n-measurable and

$$< \mu_n, |\overset{\infty}{f}| > = (n!)^{-1} < \mu , |f| >^n$$
$$< \mu_n, \overset{\infty}{f} > = (n!)^{-1} < \mu , f >^n$$

hence $\overset{\infty}{f}$ is $S\mu$-measurable and we have

$$< S\mu , |\overset{\infty}{f}| > = \sum_n (n!)^{-1} < \mu , |f| >^n = \exp(< \mu , f >)$$

and similarly (1.6).

In particular for every μ-measurable real f we have

$$< S\mu , \exp(i\overset{\circ}{f}) > \ = \ \exp(< \mu \ , e^{if} >) \qquad (1.7)$$

since $(e^{if})^{oo} = e^{i\overset{\circ}{f}}$.

(iv) For any $f,g \in L^2(X,\mu)$ we have $\overset{oo}{f}, \overset{oo}{g} \in L^2(SX,S\mu)$ and

$$< S\mu , \overset{oo}{f}\ \overline{\overset{oo}{g}} > \ = \ \exp(< \mu \ , f\ \bar{g} >). \qquad (1.8)$$

In fact $\overset{oo}{f}\ \overline{\overset{oo}{g}} = (f\ \bar{g})^{oo}$ and we can apply property (iii).

(v) The functions $\overset{oo}{f}$, with f μ-measurable and of modulus 1, are total in $L^2(SX,S\mu)$.

<u>Proof</u>. Since X is standard its Borel structure is defined by some topology making X compact and separable ; then for each m, the set $S^oX \cup \dots \cup S^mX$ is also compact and separable, and the restriction of $S\mu$ to this subset is a Radon measure. Denote by Γ the set of all continuous f on X having modulus 1 ; $\overset{oo}{\Gamma} = \{ \overset{oo}{f} \mid f \in \Gamma \}$ is closed under multiplication and conjugation, it contains the constant 1 ; moreover it separates the points in SX. In fact in the N-picture we have

$$\overset{oo}{f}(N) = \prod_{x \in X} f(x)^{N(x)} \quad ;$$

suppose $N \neq N'$; there exist $x_o, x_1, \dots x_r$ such that

$$N(x) = N'(x) = 0 \quad \text{if} \quad x \neq x_o, x_1, \dots x_r$$

$$N(x_o) \neq N'(x_o) \ ;$$

there exists $f \in \Gamma$ such that $f(x_1) = \dots = f(x_r) = 1$ and $f(x_o)$ is not a root of 1 ; then

$$\overset{oo}{f}(N) = f(x_o)^{N(x_o)} \neq f(x_o)^{N'(x_o)} = \overset{oo}{f}(N') .$$

By Stone-Weierstrass theorem the linear combinations of elements of

$\overset{\circ\circ}{\Gamma}$, when restricted to $S^oX \cup \ldots \cup S^mX$, are uniformly dense in the continuous functions on this subset. Now to prove our assertion it is sufficient to approximate in the L^2 norm an arbitrary function in $L^2(SX, S\mu)$ by a function concentrated on some subset $S^oX \cup \ldots \cup S^mX$.

(vi) For every $f \in L^1(X,\mu)$ we have $\overset{\circ}{f} \in L^1(SX, S\mu)$ and

$$< S\mu, \overset{\circ}{f} > = e^{\|\mu\|} <\mu, f> \quad . \tag{1.9}$$

In fact for every n, $\overset{\circ}{f}$ is μ_n-measurable and

$$\begin{aligned} <\mu_n, |\overset{\circ}{f}| > &= (n!)^{-1} \int | f(x_1) + \ldots + f(x_n) | \, d\mu(x_1) \ldots d\mu(x_n) \\ &\leq (n!)^{-1} \int (|f(x_1)| + \ldots + |f(x_n)|) \, d\mu(x_1) \ldots d\mu(x_n) \\ &= (n!)^{-1} \, n <\mu, |f|> \cdot \|\mu\|^{n-1} \end{aligned}$$

$$\begin{aligned} < S\mu, |\overset{\circ}{f}| > &= \sum_{n=1}^{\infty} ((n-1)!)^{-1} <\mu, |f|> \cdot \|\mu\|^{n-1} \\ &= e^{\|\mu\|} <\mu, |f|> \end{aligned}$$

which proves that $\overset{\circ}{f} \in L^1(SX, S\mu)$. A similar calculation yields (1.9).

(vii) For every $f, g \in L^2(X,\mu)$ we have $\overset{\circ}{f}, \overset{\circ}{g} \in L^2(SX, S\mu)$ and

$$< S\mu, \overset{\circ}{f} \, \overline{\overset{\circ}{g}} > = e^{\|\mu\|} (<\mu, f\bar{g}> + <\mu, f> . <\mu, \bar{g}>) \tag{1.10}$$

The proof is similar to that of (vi).

(viii) Suppose that X is a Borel sum $\sum X_i$ and set $\mu_i = \mu | X_i$; then the canonical bijection $T : SX \longrightarrow \Pi' SX_i$ carries $S\mu$ into the restricted product of the $S\mu_i$'s (see def.in App.D.3).

Proof. We recall that T carries every element $\sum_{i,n} x_{i,n}$ where $x_{i,n} \in X_i$ into the family $(\sum_n x_{i,n})_{i\in I}$; for each finite subset J of I we set $X^{(J)} = \bigcup_{i\in J} X_i \subset X$; $SX^{(J)}$ can be considered as a subset of SX, namely the subset of all $\sum_{i,n} x_{i,n}$ where i runs over J ; SX is the union of these subsets. On the other hand T carries $SX^{(J)}$ onto $\prod_{i\in J} SX_i \times \prod_{i\in I-J} \{\omega_i\} \subset \prod'_{i\in I} SX_i$ where ω_i is the zero element of SX_i. It is sufficient to prove that T carries $S\mu \mid SX^{(J)}$ into $\nu \mid \prod_{i\in J} SX_i \times \prod_{i\in I-J} \{\omega_i\}$ where ν is the restricted product of the $S\mu_i$'s. In other words it is sufficient to prove our assertion in the case where I is finite ; by induction we can suppose that I has two elements.

Then we have two measure spaces $(X,\mu),(Y,\nu)$ and their disjoint union $(X\cup Y,\mu+\nu)$. The set $(X\cup Y)^n$ can be written as $\tilde{\bigcup}_{p=0} Z_{n,p}$ where $Z_{n,p}$ is the union of the products containing p times X and n-p times Y ; there are $\binom{n}{p}$ such products and we can write

$$Z_{n,p} = \bigcup_q Z_{n,p,q} \qquad \text{with } q = 1,\dots. \binom{n}{p} .$$

Similarly we can write

$$(\mu+\nu)^{\otimes n} = \sum_{p=0} \sum_q \pi_{n,p,q}$$

where $\pi_{n,p,q}$ is a measure on $Z_{n,p,q}$. Now dropping the letter T we have

$$S^n(X\cup Y) = \tilde{\bigcup}_{p=0} S^pX \times S^{n-p}Y \qquad \text{(disjoint union) ;}$$

the canonical mapping $P_n : (X \cup Y)^n \longrightarrow S^n(X \cup Y)$ carries, for fixed n and p, all $Z_{n,p,q}$ onto $S^pX \times S^{n-p}Y$; hence

$$n!\ S(\mu+\nu) \mid S^n(X\cup Y) = P_n((\mu+\nu)^{\otimes n})$$
$$= \sum_{p=o}^{n} \sum_{q} P_n(\pi_{n,p,q})$$
$$= \sum_{p=o}^{n} \binom{n}{p} P_n(\mu^{\otimes p} \otimes \nu^{\otimes n-p})$$
$$= \sum_{p=o} \binom{n}{p} p!\ S\mu \mid S^p X \otimes (n-p)!\ S\nu \mid S^{n-p}Y$$
$$= n!\ S\mu \otimes S\nu \mid S^n(X\cup Y)$$

which proves that $S(\mu+\nu) = S\mu \otimes S\nu$.

§ 1.5. Application to linear processes and factored probability spaces

This paragraph will not be used before chapt.8. We show now the notion of symmetric measure space allows to construct factored probability spaces in the sense of [11], which are generated by decomposable linear processes.

Definition 1.6. Let T be a standard Borel space with Borel structure $\mathcal{A}$, and $(\Omega, \mathcal{B}, P)$ a probability space such that $(\Omega, \mathcal{B})$ is also standard ; we say that $(\Omega, \mathcal{B}, P)$ is a factored probability space indexed by $\mathcal{A}$ if we are given for every $A \in \mathcal{A}$ a sub-σ-algebra $\mathcal{B}_A$ of $\mathcal{B}$ with the following properties :

(i) $\mathcal{B}_T = \mathcal{B}$

(ii) $\mathcal{B}_A$ contains all P-negligible subsets

(iii) if $A_1, \dots A_n$ are mutually disjoint, $\mathcal{B}_{A_1}, \dots \mathcal{B}_{A_n}$ are independent

(iv) if $A_1, A_2, \ldots$ are mutually disjoint with union A, $\mathcal{B}_A$ is generated up to P-negligible subsets by $\mathcal{B}_{A_1}, \mathcal{B}_{A_2}, \ldots$

We now give examples of factored probability spaces. We consider a standard Borel space X with a finite positive Borel measure μ, a standard Borel space $(T, \mathcal{A})$ and a Borel mapping p of X into T ; we set

$$\Omega = SX \qquad\qquad P = S\mu / e^{\|\mu\|} \quad ;$$

we denote by $\mathcal{B}$ the σ-algebra on Ω generated by the Borel structure of SX (see § 1.3) and all P-negligible subsets ; for every $A \in \mathcal{A}$, X is a disjoint union $p^{-1}(A) \cup p^{-1}(T-A)$ and by § 1.3 we can write

$$\Omega = S(p^{-1}(A)) \times S(p^{-1}(T-A)) \quad ;$$

we denote by $\mathcal{B}_A$ the sub-σ-algebra of $\mathcal{B}$ generated by all P-negligible subsets and all subsets of the form $B \times S(p^{-1}(T-A))$ where B is a Borel subset of $S(p^{-1}(A))$. Then conditions (i) and (ii) above are trivially satisfied ; condition (iii) follows from the fact that if X is a disjoint union $X = A_1 \cup \ldots \cup A_n$, the measure space $(SX, S\mu)$ is the product of the measure spaces $(SA_i, S(\mu/A_i))$; finally (iv) follows from proposition D.1. In § 8.5 we shall look more closely at this type of factored probability spaces.

We now pass to the construction of certain Poissonian decomposable linear processes (see definition in App.E.3) associated with our factored probability space. We denote by E the vector space of all real Borel functions on T taking only a finite number of values ; by [37] § 0 the measure space (X, μ) is isomorphic to a measure space of the form $(T \times \underline{R}, \pi)$ where π is some positive finite measure on $T \times \underline{R}$, in such a manner that p becomes the projection $T \times \underline{R} \longrightarrow T$; let ω be a real Borel function on $\underline{R}$. For any $f \in E$ we construct the function $f \otimes \omega$ on $T \times \underline{R}$:

$$(f \otimes \omega)(t,a) = f(t)\,\omega(a)$$

and then the function $(f \otimes \omega)^{o}$ on $S(T \times \underline{R})$; we set

$$X_f = (f \otimes \omega)^{o} ;$$

we thus get a decomposable linear process ; its characteristic functional is given by

$$\begin{aligned} \varphi(f) &= \langle P, \exp(iX_f) \rangle \\ &= e^{-\|\pi\|} \langle S\pi, \exp(i(f \otimes \omega)^{o}) \rangle \ ; \end{aligned}$$

by (1.7)

$$\begin{aligned} \varphi(f) &= e^{-\|\pi\|} \exp(\langle \pi, e^{i\, f \otimes \omega} \rangle) \\ &= \exp\left[\int (e^{if(t)\,\omega(a)} - 1)\, d\pi(t,a)\right] . \end{aligned} \qquad (1.11)$$

This can be written $\exp\left[\int (e^{i\lambda f(t)} - 1)\, d\gamma(t,\lambda)\right]$ where γ is the image of π under the mapping $(t,a) \longmapsto (t, \omega(a))$; thus we see that our linear process is Poissonian. Using the N-picture we can write X_f as follows : for every $N \in S(T \times \underline{R})$ we have

$$X_f(N) = \sum_{t,a} f(t)\,\omega(a)\, N(t,a). \qquad (1.12)$$

Moreover the linear process $f \longmapsto X_f$ is <u>measurable with respect to the factored probability space</u> (see [11]), which means that for every $A \in \mathcal{A}$, X_f is measurable with respect to $\mathcal{B}_A$ if f is null outside of A.

We now prove that the factored probability space is <u>lineariza-</u>

ble (see [11]), i.e. that there exist decomposable linear processes $X^{(1)}$; $X^{(2)}$,... which are measurable with respect to the factored probability space and have the property that each $\mathcal{B}_A$ is generated up to P-negligible sets by the functions $X_B^{(n)}$ with $B \in \mathcal{a}$, $B \subset A$ (we write X_B instead of X_{χ_B}). To this aim we take real Borel functions on $\underline{R}$: $\omega_1, \omega_2, \ldots$ with the following property : if f is a real function on $\underline{R}$ which is null outside of a finite subset, then

$$\sum_{a \in \underline{R}} f(a)\,\omega_n(a) = 0 \quad \forall n \implies f = 0$$

[for instance $\omega_n = \chi_{U_n}$ where (U_n) is a basis of open subsets in $\underline{R}$]; we set

$$X_f^{(n)} = (f \otimes \omega_n)^o .$$

Then denoting by $\mathcal{B}'_A$ the smallest sub-σ-algebra of $\mathcal{B}$ containing all P-negligible sets and making measurable all functions $X_B^{(n)}$, $B \in A$, we have to prove the following

Proposition 1.1. For every $A \in \mathcal{a}$ we have $\mathcal{B}'_A = \mathcal{B}_A$.

We already know that $\mathcal{B}'_A \subset \mathcal{B}_A$ and we must prove the converse inclusion. $\mathcal{B}_A$ and $\mathcal{B}'_A$ are inverse images under the projection $S(T \times \underline{R}) \longrightarrow S(A \times \underline{R})$ of two Borel structures on $S(A \times R)$; the first one is standard, the other one is less fine ; it is sufficient to prove that the second one is generated by a sequence of subsets which moreover separate the points (see [9], App.B). To do this we can suppose $A = T$; then $\mathcal{a}$ is generated by a sequence $A_1, A_2, \ldots$ and we can suppose that the set $\{A_n\}$ contains T and is closed for finite unions and complementation ; then it is known (see e.g.[17]lemma 6.1) that $\mathcal{A}$ is the smallest set containing $\{A_m\}$ and closed for countable increasing unions and complementation ; but if $A_{m_i} \nearrow A$, $X^{(n)}_{A_{m_i}}$ converges simply to $X^{(n)}_A$, and on the other hand we have

$$X^{(n)}_{T-A_m} = X^{(n)}_T - X^{(n)}_{A_m} .$$

This proves that the functions $X^{(n)}_{A_m}$ generate $\mathcal{B}'$; now take a countable basis of open subsets U_p in $\underline{R}$; the subsets

$$V_{m,n,p} = (X^{(n)}_{A_m})^{-1}(U_p)$$

generate $\mathcal{B}'$. Let us now prove that they separate the points of $S(T\times \underline{R})$.

Let N,N' two distinct elements in $S(T\times \underline{R})$; there exists a finite set $t_o,\dots t_q$ in T such that

$$N(t,a) = N'(t,a) = 0 \quad \text{if} \quad t \neq t_o,\dots t_q ;$$

since $N \neq N'$ we can suppose that there exists a_o such that

$$N(t_o,a_o) \neq N'(t_o,a_o) ;$$

take m such that A_m contains t_o but not $t_1,\dots t_q$; there exists n such that

$$\sum_a \omega_n(a)\, N(t_o,a) \neq \sum_a \omega_n(a)\, N'(t_o,a) ;$$

there exists p such that

$$\sum_a \omega_n(a)\, N(t_o,a) \in U_p$$
$$\sum_a \omega_n(a)\, N'(t_o,a) \notin U_p ;$$

then by (1.12) we have

$$X_{A_m}^{(n)}(N) = \sum_a \omega_n(a)\, N(t_o,a) \in U_p$$

$$X_{A_m}^{(n)}(N') = \sum_a \omega_n(a)\, N'(t_o,a) \notin U_p$$

i.e.

$$N \in V_{m,n,p} \qquad N' \in V_{m,n,p} \ .$$

Chapter 2. THE SYMMETRIC HILBERT SPACE OF A HILBERT SPACE

In this chapter we introduce the symmetric Hilbert spaces as well as the vectors EXP a and the unitary operators $U_{A,b,c}$ which will play a fundamental role throughout this lecture.

§ 2.1. Definitions and general properties

Given a Hilbert space H over the field $K = \underline{R}$ or $\underline{C}$ we can form the n-th Hilbert tensor power $H^{\otimes n}$ where $n = 1,2,..$; the symmetric group $\mathfrak{S}_n$ acts by unitary operators $U_{s,n}$ in $H^{\otimes n}$: for each $s \in \mathfrak{S}_n$ we have

$$U_{s,n}(x_1 \otimes \dots \otimes x_n) = x_{s(1)} \otimes \dots \otimes x_{s(n)} \quad ;$$

the operator $P_n = (n!)^{-1} \sum_{s \in \mathfrak{S}_n} U_{s,n}$ is an orthogonal projection, the image of which is denoted by S^nH, the set of all elements in $H^{\otimes n}$ which are invariant by all $U_{s,n}$, and called the n-th symmetric power of H. We set $S^oH = K$; we have $S^1H = H$. One gets an orthonormal basis of S^nH in the following way : let $e_1, e_2, ..$ be an orthonormal basis of H (we suppose H separable for the sake of simplicity); for any sequence of positive integers $n_1, n_2, ...$ such that $n_1 + n_2 + ... = n$, the element

$$(n!)^{1/2}(n_1!\ n_2!\ \dots)^{-1/2} P_n(e_1^{\otimes n_1} \otimes e_2^{\otimes n_2} \otimes \dots)$$

belongs to S^nH ; and these elements constitute an orthonormal basis of S^nH.

Definition 2.1. The symmetric Hilbert space SH of H is the Hilbert sum $\bigoplus_{n=0}^{\infty} S^n H$, that is the set of all sequences $x = (x_n)$ with $x_n \in S^n H$ and $\sum \| x_n \|^2 < \infty$; the scalar product is given by $(x \mid y) = \sum (x_n \mid y_n)$.

The space SH is sometimes called "exponential space of H" (see [2] and [25]) ; we shall give in example 3.1 a new interpretation of SH (for a real H) as the real Hilbert space associated with a kernel on H.

Definition 2.2. For every $a \in H$ we denote by EXP a the element $x = (x_n) \in SH$ such that $x_n = (n!)^{-1/2} a^{\otimes n}$, that is

$$\text{EXP } a = (1 , a , (2!)^{-1/2} a \otimes a , (3!)^{-1/2} a \otimes a \otimes a , \dots) ;$$

we have

$$(\text{EXP } a \mid \text{EXP } b) = e^{(a \mid b)}$$

$$\| \text{EXP } a \| = e^{\| a \|^2 / 2} \geq \| a \|$$

$$| (\text{EXP } a \mid \text{EXP } b) | = e^{\text{Re}(a \mid b)} .$$

The element EXP 0 will play a central role in the theory. The vectors EXP a have been introduced for physical purpose by Araki and Woods in [2] and independently, for probabilistic purpose, by G. Kallianpur [24] and J.Neveu [30]; they are used in theoretical physics under the name of "coherent states" (see [36]).

Proposition 2.1. The mapping EXP of H into SH is injective and bicontinuous.

It is continuous since if a_n tends to a we have

$$\| \mathrm{EXP}\ a_n - \mathrm{EXP}\ a\|^2 = \| \mathrm{EXP}\ a_n\|^2 + \| \mathrm{EXP}\ a\|^2 - 2\mathrm{Re}(\mathrm{EXP}\ a_n | \mathrm{EXP}\ a)$$

$$= \exp(\| a_n\|^2) + \exp(\| a\|^2) - 2\ \mathrm{Re}\ \exp(a_n | a)$$

which tends to 0. It is injective and bicontinuous since the second coordinate of EXP a is equal to a.

<u>Proposition</u> 2.2. The elements EXP a , $a \in H$, are linearly independent and total in SH.

a) We first prove that if $c_1, .. c_q$ are non zero vectors in H there exists a vector b in H such that $(c_i | b) \neq 0 \quad \forall\ i = 1, .. q$. We can suppose that the c_i's are mutually non proportional. Take an arbitrary vector b_o ; it can be orthogonal to some of the c_i's, say to $c_1, .. c_r$, and non orthogonal to $c_{r+1}, .. c_q$; we can find a vector b_1 non orthogonal to c_1 and sufficiently near to b_o to be non orthogonal to $c_{r+1}, .. c_q$; and so on ; finally we obtain a vector which is non orthogonal to $c_1, .. c_q$.

b) We now prove that the EXP a are linearly independent. Suppose we have distinct vectors $a_1, .. a_n$ and non zero scalars $\lambda_1, .. \lambda_n$ such that $\sum_{i=1}^{n} \lambda_i\ \mathrm{EXP}\ a_i = 0$; for each $x \in H$ we have

$$f(x) = (\sum \lambda_i\ \mathrm{EXP}\ a_i | \mathrm{EXP}\ x)$$

$$= \sum \lambda_i\ e^{(a_i | x)} = 0 ;$$

consider f as a mapping of the real vector space H into the real vector space K ; its p-th derivative at a point x is given by

$$f_x^{(p)}(b_1, .. b_p) = \sum_{i=1}^{n} \lambda_i\ e^{(a_i | x)}\ (a_i | b_1) \dots (a_i | b_p)$$

and will be identically zero ; in particular for each $b \in H$ we shall have

$$f_o^{(p)}(b,..\, b) = \sum_i \lambda_i (a_i|b)^p = 0 \quad \forall\, p = o,1,..\, n-1 ;$$

this is a linear system with respect to the λ_i's which admits a non trivial solution ; its determinant (of Vandermonde type) must be zero, and there exist two distinct indices i and j such that $(a_i|b) = (a_j|b)$, i.e. $(a_i - a_j|b) = 0$; the vectors $a_i - a_j$ for $i \neq j$ are not null and every vector b is orthogonal to at least one of them, which is impossible by part a) of the proof.

c) We finally prove that the EXP a are total in SH, i.e. that the closed linear subspace L generated by them is equal to SH. Since S^nH is generated by tensors of the form $a^{\otimes n}$, it is sufficient to prove that each $a^{\otimes n}$ belongs to L. For $t \in \underline{R}$ set $f(t) = \text{EXP}\ ta$; we have $f^{(n)}(0) = (n!)^{\frac{1}{2}} a^{\otimes n}$; since $f(t) \in L\ \forall t$ we have $f^{(n)}(t) \in L$ and $a^{\otimes n} \in L$.

<u>Example</u> 2.1. The space SK, with $K \in \underline{R}$ or C, is canonically isomorphic to $L^2(\underline{R}, \nu ;K)$ where ν is the reduced Gaussian measure on $\underline{R}$. In fact by App.E.1 the functions $(n!)^{-\frac{1}{2}} h_n$ form an orthonormal basis of $L^2(\underline{R}, \nu ;K)$; since the elements $1^{\otimes n}$ form an orthonormal basis of SK, there exists an isomorphism $SK \longrightarrow L^2(\underline{R}, \nu ;K)$ which carries every $1^{\otimes n}$ into $(n!)^{-\frac{1}{2}} h_n$ and every EXP a into the function

$$x \longmapsto \sum_{n=o}^{\infty} (n!)^{-1} a^n h_n(x) = e^{ax-a^2/2} .$$

Proposition 2.3. Suppose H is the Hilbert sum of a family of Hilbert spaces $(H_i)_{i\in I}$; denote by u_i the element EXP 0 of SH_i. There exists a unique isomorphism $T : SH \longrightarrow \bigotimes_{i\in I}^{h(u_i)} SH_i$ carrying EXP a into $\otimes$ EXP a_i for every element $a = (a_i) \in H$ with $a_i = 0$ a.e.

(For the notion of infinite tensor product of Hilbert spaces, see App.A).

Consider $a = (a_i)$, $b = (b_i)$ with $a_i = b_i = 0$ a.e.; we have

$$(\text{EXP } a | \text{EXP } b) = e^{(a|b)} = e^{\Sigma (a_i|b_i)} = \prod e^{(a_i|b_i)}$$

$$= (\otimes \text{ EXP } a_i | \otimes \text{ EXP } b_i) \ ;$$

then the proposition follows from the fact that the elements EXP a are total in SH and the elements $\otimes$ EXP a_i are total in $\bigotimes^{h(u_i)} SH_i$.

Remark 2.0. Denote by E_c the complexification of a real vector space E ; then SH_c is canonically isomorphic to $(SH)_c$. In fact consider an element $x+iy$ in H_c with $x,y \in H$; developing the element $(x+iy)^{\otimes n}$ we get first a number of members containing an even power of y and secondly a number of members containing an odd power of y ; denote respectively by a and ib the sum of these members ; then there is a unique isomorphism of S^nH_c onto $(S^nH)_c$ carrying $(x+iy)^{\otimes n}$ into $a+ib$.

§ 2.2. The unitary operators $U_{A,b,c}$ and the group $\mathcal{G}_H$ —

We denote by H a complex Hilbert space and by H^1 the set of all elements in SH of the form λ EXP a with $\lambda \in \underline{C}^* = \underline{C} - \{0\}$ and

$a \in H$; we recall that $\lambda \, EXP\, a = \lambda' EXP\, a'$ implies $\lambda = \lambda'$, $a = a'$.

Definition 2.3. We denote by $\mathcal{G}_H$ the group of all unitary operators U in SH such that U and U^{-1} preserve H^1 globally. Our next aim is to determine the general form of these operators. For any Hilbert space H, $\mathcal{U}(H)$ will denote the group of all unitary operators in H.

First each unitary operator A in H extends in a natural way to a unitary operator in SH, namely $(I, A, A^{\otimes 2}, \ldots, A^{\otimes n}, \ldots)$; we call it $U_{A,0,1}$; we thus have

$$U_{A,0,1}\,(EXP\; x) = EXP\; Ax \tag{2.1}$$

and therefore $U_{A,0,1}$ belongs to $\mathcal{G}_H$; moreover the mapping $A \longrightarrow U_{A,0,1}$ is a morphism of groups.

Secondly consider a vector b in H ; for each x and y in H we have

$$(EXP(x+b) \mid EXP(y+b)) = \exp\left[(x|y) + (x|b) + (b|y) + \|b\|^2\right];$$

setting

$$X = \exp\left(-\tfrac{1}{2}\|b\|^2 - (x|b)\right) EXP\,(x+b)$$

$$Y = \exp\left(-\tfrac{1}{2}\|b\|^2 - (y|b)\right) EXP\,(y+b)$$

we have

$$(X \mid Y) = e^{(x|y)} = (EXP\; x \mid EXP\; y)\;;$$

since the elements EXP x and X (b being fixed) are total in SH, there exists a unique unitary operator in SH carrying each EXP x into the corresponding X ; we call it $U_{I,b,1}$ and we have

$$U_{I,b,1}\,(EXP\; x) = \exp\left(-\tfrac{1}{2}\|b\|^2 - (x|b)\right) EXP(x+b) \tag{2.2}$$

and moreover $U_{I,b,1}$ belongs to $\mathcal{G}_H$. One easily checks that

$$U_{I,b,1}\; U_{I,b',1} \;=\; \exp(i\,\mathrm{Im}(b|b'))\, U_{I,b+b',1}. \qquad (2.3)$$

Now for every $c \in \underline{U}$ (the group of all complex numbers having modulus one), every $b \in H$ and every $A \in \mathcal{U}(H)$ we set

$$U_{A,b,c} \;=\; c\, U_{I,b,1}\, U_{A,0,1} \;;$$

clearly $U_{A,b,c}$ belongs to $\mathcal{G}_H$; by (2.1) and (2.2) we have

$$U_{A,b,c}\,(\mathrm{EXP}\; x) \;=\; c\,\exp(-\|b\|^2/2 - (Ax|b))\,\mathrm{EXP}(Ax + b) \qquad (2.4)$$

in particular

$$U_{A,b,c}\,(\mathrm{EXP}\; 0) \;=\; c\,\exp(-\|b\|^2/2)\,\mathrm{EXP}\; b \qquad (2.5)$$

moreover one easily checks that

$$U_{A,b,c}\; U_{A',b',c'} \;=\; U_{AA',\,b+Ab',\,cc'\exp(i\,\mathrm{Im}(b|Ab'))} \qquad (2.6)$$

<u>Lemma</u> 2.1. The mapping $(A,b,c) \longmapsto U_{A,b,c}$ from $\mathcal{U}(H) \times H \times \underline{U}$ into $\mathcal{G}_H$ is bijective.

Proof of the injectivity : suppose $U_{A,b,c} = U_{A',b',c'}$; (2.5) implies $b = b'$ and $c = c'$; then (2.4) implies $Ax+b = A'x+b$ for every x, i.e. $A = A'$.

Proof of the surjectivity. Let U be an arbitrary element of $\mathcal{G}_H$; $U(\mathrm{EXP}\; 0)$ is of the form $\lambda\,\mathrm{EXP}(-b)$ with $\lambda \in \underline{C}^*$, $b \in H$; multiplying U by a scalar of modulus one we can suppose $\lambda > 0$; we then have

$$U_{I,b,1}\,(U(\mathrm{EXP}\; 0)) \;=\; \lambda\,\exp(-\|b\|^2/2 + \|b\|^2)\,\mathrm{EXP}\; 0$$
$$= \lambda\,\exp(\|b\|^2/2)\,\mathrm{EXP}\; 0 \;;$$

since U is unitary and $\lambda \exp(\|b\|^2/2) > 0$ we have $\lambda \exp(\|b\|^2/2) = 1$; setting $U_{I,b,1}\, U = U'$ we have $U' \in \mathcal{E}_H$ and $U'(\mathrm{EXP}\ 0) = \mathrm{EXP}\ 0$. We are thus led to prove our assertion in the case where $U(\mathrm{EXP}\ 0) = \mathrm{EXP}\ 0$. For every x in H, U(EXP x) is of the form $\lambda_x\ \mathrm{EXP}(Tx)$ where T is a bijection of H onto itself ; we have

$$1 = (\mathrm{EXP}\ x \mid \mathrm{EXP}\ 0) = (U(\mathrm{EXP}\ x) \mid U(\mathrm{EXP}\ 0))$$

$$= \lambda_x\, (\mathrm{EXP}\ Tx \mid \mathrm{EXP}\ 0) = \lambda_x \ ;$$

for x,y in H

$$e^{(x|y)} = (\mathrm{EXP}\ x \mid \mathrm{EXP}\ y) = (U(\mathrm{EXP}\ x) \mid U(\mathrm{EXP}\ y))$$

$$= (\mathrm{EXP}\ Tx \mid \mathrm{EXP}\ Ty) = e^{(Tx|Ty)}$$

$$(Tx \mid Ty) = (x \mid y) \mod 2\pi i\, \underline{Z} \qquad (2.7)$$

$$\mathrm{Re}\,(Tx \mid Ty) = \mathrm{Re}\,(x \mid y) \ .$$

Let us consider H as a real Hilbert space with scalar product Re (x | y) ; T is a bijection of H which preserves 0 and the scalar products ; therefore it is an $\underline{R}$-linear and orthogonal operator. Now by (2.7) for any real λ we have

$$\lambda((Tx \mid Ty) - (x|y)) = (Tx \mid T\lambda y) - (x \mid \lambda y) \in 2\pi i\, \underline{Z}$$

which implies $(Tx|Ty) - (x|y) = 0$; for the complex structure of H, T is a bijection preserving 0 and the scalar products ; therefore it is a $\underline{C}$-linear and unitary operator. Finally $U = U_{T,0,1}$.

<u>Theorem</u> 2.1. The set $\mathcal{U}(H) \times H \times \underline{U}$ is a group for the following composition law :

$$(A,b,c)\,(A',b',c') = (AA',b+Ab',cc' \exp(i\ \mathrm{Im}(b|Ab'))) \qquad (2.8)$$

with neutral element $(I,0,1)$; it is a topological group for the product topology of the strong topology on $\mathcal{U}(H)$, the strong topology on H and the ordinary topology on $\underline{U}$. The mapping $(A,b,c) \longmapsto U_{A,b,c}$ is a bicontinuous isomorphism of $\mathcal{U}(H) \times H \times \underline{U}$ onto $\mathcal{G}_H$ endowed with the strong topology.

We have only to prove the bicontinuity of the mapping. If a net (A_i,b_i,c_i) converges to an element (A,b,c), in order to prove that U_{A_i,b_i,c_i} converges to $U_{A,b,c}$ it is sufficient to prove that $U_{A_i,b_i,c_i}(X) \longrightarrow U_{A,b,c}(X)$ for vectors X belonging to some total set, for instance of the form EXP x ; the assertion is then trivial by (2.4). Conversely suppose that U_{A_i,b_i,c_i} converges to $U_{A,b,c}$; for every x in H we have

$$c_i \exp(-\|b_i\|^2/2 -(A_ix|b_i)) \,\mathrm{EXP}(A_ix + b_i) \longrightarrow$$

$$c \exp(-\|b\|^2/2 -(Ax|b)) \,\mathrm{EXP}(Ax + b) \;;$$

taking the first coordinate of both members we see that

$$c_i \exp(-\|b_i\|^2/2 -(A_ix|b_i)) \longrightarrow c \exp(-\|b\|^2/2 -(Ax|b)) \;; \tag{2.9}$$

then taking the second coordinate we see that $A_ix + b_i$ tends to $Ax + b$ for every x, which implies $b_i \longrightarrow b$, $A_i \longrightarrow A$; finally (2.9) shows that c_i tends to c.

<u>Remark</u> 2.1. The group $\mathcal{U}(H) \times H \times \underline{U}$ has a normal subgroup $I \times 0 \times \underline{U}$ with quotient isomorphic to the group of the transformations $x \longmapsto Ax + b$, i.e. the group Γ of rigid motions in H ; the mapping $((A,b),(A',b')) \longmapsto \exp(i \,\mathrm{Im}(b|Ab'))$ is a cocycle (or a multiplier

in the terminology of [26]) on Γ and $\mathcal{U}(H) \times H \times \underline{U}$ is the extension of Γ by $\underline{U}$ determined by this cocycle.

<u>Proposition</u> 2.4. We keep the notations of prop.2.3 and consider operators U_{A_i, b_i, c_i} in SH_i with $A_i = I$, $b_i = 0$, $c_i = 1$ a.e.; we can form the operator $\otimes U_{A_i, b_i, c_i}$ in the space $\overset{h(u_i)}{\otimes} SH_i$. Then T^{-1} carries $\otimes U_{A_i, b_i, c_i}$ into $U_{A,b,c}$ where $A = \oplus A_i$, $b = \oplus b_i$, $c = \prod c_i$.

It is enough to prove that

$$T(U_{A,b,c}(\text{EXP } x)) = (\otimes U_{A_i, b_i, c_i})(T(\text{EXP } x))$$

which is an easy computation.

<u>The Fock representation</u>

The mapping $b \longmapsto U_{I,b,1}$ is a projective representation of the additive group of H in the Hilbert space SH with multiplier

$$(b,b') \longmapsto \exp(i \operatorname{Im}(b|b'))$$

(for the definition of projective representations see e.g. [26]). It is known in Quantum Field Theory as the Fock representation (see App.F).

<u>Proposition</u> 2.5. The above representation is irreducible.

We take an orthonormal basis $(e_i)_{i \in I}$ of H and denote by E the subspace algebraically generated by the e_i's ; by prop.2.3 and 2.4 we can write

$$SH = \overset{h(u_i)}{\otimes} S(\underline{C}\, e_i)$$

$$U_{I,b,1} = \otimes U_{I,b_i,1}$$

for every $b = \sum b_i e_i \in E$; by App.A.4 it is enough to prove that the representation $b_i \longmapsto U_{I,b_i,1}$ of $\underline{C}\, e_i$ in $S(\underline{C}\, e_i)$ is irreducible for every i ; thus we can assume that $H = \underline{C}$. Then we carry our representation into $L^2(\underline{R}, \nu)$ by the isomorphism T described in example 2.1 ; by an easy computation we get

$$(T.U_{I,b,1}.T^{-1}.f)(x) = \exp(-b_1^2 - ib_1 b_2 + bx).f(x - 2b_1)$$

for every $b = b_1 + ib_2 \in \underline{C}$ and every $f \in L^2(\underline{R}, \nu)$; now this representation is irreducible. In fact if an operator W in $L^2(\underline{R},\nu)$ commutes with all $T.U_{I,b,1}.T^{-1}$, it commutes with all $T.U_{I,ib,1}.T^{-1}$ with real b and therefore with all multiplication operators ; then W is itself the multiplication operator by some function g ; since W commutes with all operators $T.U_{I,b,1}.T^{-1}$ with real b, g must be a constant and W is a scalar operator.

Remark 2.2. For a real Hilbert space H we can also define operators $U_{A,b,c}$ but here $c = \pm 1$ and formula (2.8) becomes

$$(A,b,c)\ (A',b',c') = (AA', b + Ab', cc') ;$$

the mapping $b \longmapsto U_{I,b,1}$ is now an ordinary representation of H in SH ; but it is no longer irreducible.

§ 2.3. Relation with symmetric measure spaces

Proposition 2.6. Let X be a standard Borel space with a finite positive Borel measure μ ; there exists a unique isomorphism of $S(L^2(X,\mu))$ onto $L^2(SX,S\mu)$ carrying EXP f into $\overset{\infty}{f}$ for every f in $L^2(X,\mu)$. This isomorphism carries $(n!)^{-\frac{1}{2}} f^{\otimes n}$ into the function $x_1 + \dots x_n \longmapsto f(x_1)\dots f(x_n)$.

Take f,g in $L^2(X,\mu)$; by § 1.4 property (iv) we have

$$(\overset{\circ\circ}{f} \mid \overset{\circ\circ}{g}) = e^{(f|g)} = (\mathrm{EXP}\ f \mid \mathrm{EXP}\ g) ;$$

the proposition follows from the fact that the elements $\overset{\circ\circ}{f}$ and EXP f are total respectively in $L^2(SX, S\mu)$ (see § 1.4 property (v)) and $SL^2(X,\mu)$.

Remark 2.3. Proposition 2.6 can be generalized as follows : we consider a Hilbert integral $H = \int_X^{\oplus} H_x \, d\mu(x)$ (see definition in [8] or [16]); we define a field of Hilbert spaces on SX as follows : first we take a Borel total order relation $\leqslant$ on X ; now let x be an element of SX ; it can be written as

$$x = \underbrace{x_1 + \ldots x_1}_{n_1 \text{ times}} + \underbrace{x_2 + \ldots x_2}_{n_2 \text{ times}} + \ldots\ldots + \underbrace{x_r + \ldots x_r}_{n_r \text{ times}}$$

with $x_1 < x_2 < \ldots < x_r$; then we set

$$H_x = S^{n_1} H_{x_1} \overset{h}{\otimes} \ldots\ldots \overset{h}{\otimes} S^{n_r} H_{x_r} .$$

The result is : there exists a unique isomorphism of SH onto $\int_{SX}^{\oplus} H_x . dS\mu(x)$ carrying every element $(n!)^{-\frac{1}{2}} f^{\otimes n}$ with $f \in H$ into the vector $\tilde{f}$ having coordinates $\tilde{f}_x = f_{x_1}^{\otimes n_1} \otimes \ldots \otimes f_{x_r}^{\otimes n_r}$ in the above notations. This result becomes simpler when μ is non atomic ; in fact we can neglect the n-uples $(x_1, \ldots x_n)$ having at least two equal components, and define H_x almost everywhere by $H_x = H_{x_1} \overset{h}{\otimes} \ldots \overset{h}{\otimes} H_{x_n}$ if $x = x_1 + \ldots x_n$ with $x_1 < x_2 < \ldots < x_n$; then $\tilde{f}$ becomes $\tilde{f}_x = f_{x_1} \otimes \ldots \otimes f_{x_n}$.

Chapter 3. POSITIVE DEFINITE FUNCTIONS OF TYPE (S)

After a few perequisites about positive definite functions on groups we introduce a new type of positive definite functions which we call type (S) ; in the following chapter we show that they are closely related to the infinitely divisible positive definite functions.

§ 3.1. Positive definite functions

We recall that a complex matrix with coefficients $a_{i,j}$, $i,j = 1,\dots n$, is positive definite if the corresponding quadratic form is positive,i.e. if for any complex numbers $x_1,\dots x_n$ we have

$$\sum_{i,j} a_{i,j}\, x_i\, \bar{x}_j \geqslant 0 \qquad (3.0)$$

A positive definite matrix is always hermitian. If two matrices $(a_{i,j})$, $(b_{i,j})$, $i,j = 1,\dots n$, are positive definite, the matrix $(a_{i,j}\, b_{i,j})$ is also positive definite (this can be seen by using the tensor product of the two quadratic forms) ; by induction we see that the coefficientwise product of a finite number of positive definite matrices is again a positive definite matrix. It follows that if $(a_{i,j})$ is positive definite, $(e^{a_{i,j}})$ is positive definite too.

In this lecture we call kernel on a set E a complex function on $E \times E$.

Definition 3.1. A kernel K on a set E is positive definite if for any $x_1, \ldots x_n$ in E the matrix with coefficients $K(x_i, x_j)$ is positive definite. Every positive definite kernel is hermitian, i.e. $K(x,y) = \overline{K(y,x)}$.

With every positive definite kernel K one can associate canonically a Hilbert space in the following manner : we take $\underline{C}^{(E)}$, the vector space of all complex functions f on E such that $f(x) = 0$ a.e. (i.e. except for a finite set of x), we endow it with a (possibly degenerate) scalar product

$$(f|f') = \sum_{x,x'} K(x,x')\, f(x)\, \overline{f'(x')} \ ;$$

in particular

$$(\delta_x | \delta_y) = K(x,y)$$

where δ_x is the function taking the value 1 at x and 0 elsewhere (Dirac function at x). Dividing $\underline{C}^{(E)}$ by the subspace of all vectors satisfying $(f|f) = 0$ we get a separated prehilbert space ; and finally we take its completion. This Hilbert space is called the Hilbert space canonically associated with the kernel K. Replacing $\underline{C}^{(E)}$ by $\underline{R}^{(E)}$ we could also construct a real Hilbert space associated with a real kernel K.

Definition 3.2. A complex function φ on a group G is said to be positive definite if the kernel $K(g,g') = \varphi(g'^{-1}g)$ is positive definite, i.e. if for any complex numbers $c_1, \ldots c_n$ and any $g_1, \ldots g_n$ in G we have $\sum_{p,q} c_p \overline{c_q}\, \varphi(g_q^{-1} g_p) \geq 0$.

This implies that φ is hermitian, i.e. $\varphi(g^{-1}) = \overline{\varphi(g)}$, and that $|\varphi(g)| \leq \varphi(e)$ $\forall g$ where e is the neutral element of G ; is said to be normalized if $\varphi(e) = 1$. The ordinary product of a finite family of p.d.f. (positive definite functions) is again a p.d.f. ; if φ is a p.d.f., e^{φ} is also a p.d.f.

If U is a unitary representation of G in a complex Hilbert space H and if ξ is a vector in H, the function

$$g \longmapsto (U(g).\xi | \xi)$$

is positive definite ; if moreover G is a topological group and U is continuous, then φ is continuous too. If two triples (H,U,ξ) and (H',U',ξ') with ξ and ξ' cyclic define the same p.d.f., there exists an isomorphism $H \longrightarrow H'$ carrying U into U' and ξ into ξ' (we recall that a vector ξ is cyclic for a representation U if the vectors $U(g)\xi$ are total in H).

Conversely with every normalized p.d.f. φ on G one can associate canonically a complex Hilbert space H_φ, a unitary representation U_φ of G in H and a cyclic unit vector ξ_φ such that $\varphi(g) = (U_\varphi(g).\xi_\varphi | \xi_\varphi)$: first H_φ is the Hilbert space canonically associated with the kernel K ; then we denote by Λ_φ the canonical mapping $\underline{C}^{(G)} \longrightarrow H_\varphi$; we set $\xi_\varphi = \Lambda_\varphi(\delta_e)$; finally to define $U_\varphi(g)$ we first define a linear operator $U^o(g)$ in $\underline{C}^{(G)}$ by

$$(U^o(g).f)(g') = f(g^{-1}g')$$

and $U_\varphi(g)$ is nothing but the operator in H associated with $U^o(g)$. If moreover φ is continuous, U_φ is also continuous ; in fact we have to prove that if g tends to e, $(U_\varphi \eta_1 | \eta_2)$ tends to $(\eta_1|\eta_2)$ for any η_1, η_2 in H_φ ; since $U_\varphi(g)$ is unitary we can take η_1

and η_2 in some total subset, for instance $\eta_i = \Lambda_\varphi(\delta_{g_i})$; we then have

$$(U_\varphi(g)\eta_1 | \eta_2) = (\Lambda_\varphi(\delta_{gg_1}) | \Lambda_\varphi(\delta_{g_2}))$$

$$= \varphi(g_2^{-1} g\, g_1)$$

which actually tends to $\varphi(g_2^{-1} g_1) = (\eta_1 | \eta_2)$.

The above construction is called Gelfand-Segal construction. If φ is real we can also construct a real Hilbert space H_φ^r , a representation U_φ^r of G in H_φ^r and a cyclic vector ξ_φ^r ; H_φ is just the complexification of H_φ^r .

Remark 3.1. If $\varphi(g) = (U(g)\xi | \xi)$ then $e^{\varphi(g)} = (U_{U(g),0,1}$ EXP ξ | EXP $\xi)$ where $U_{A,b,c}$ is the notation introduced in § 2.2.

Example 3.1. Let E be a real separated prehilbert space, $\bar{E}$ its completion ; by remark 2.2 we have a representation U of E in $S\bar{E}$: $U(x) = U_{I,x,1}$; in particular

$$U(x).\mathrm{EXP}\ 0 = \exp(-\|x\|^2/2)\ \mathrm{EXP}\ x$$

which proves that EXP 0 is cyclic for U ; on the other hand the function

$$\varphi(x) = (U(x).\mathrm{EXP}\ 0 \mid \mathrm{EXP}\ 0) = \exp(-\|x\|^2/2)$$

is positive definite. We can thus identify H_φ^r with $S\bar{E}$, ξ_φ^r with EXP 0 , $U_\varphi^r(x)$ with U(x) and $\Lambda_\varphi^r(\delta_x) = U_\varphi^r(x)$. ξ_φ^r with the vector $\exp(-\|x\|^2/2)$.EXP x. We see that $S\bar{E}$ is nothing but the real Hilbert space associated with the kernel

$$(x,y) \longmapsto \exp(-\|x-y\|^2/2) = \exp(-\|x\|^2/2 - \|y\|^2/2 - (x|y)).$$

Relation between Gelfand-Segal construction and direct products

Lemma 3.1. Let $(G_i)_{i\in I}$ be an arbitrary family of groups and G be a subgroup of $\prod_{i\in I} G_i$ which contains the restricted product $\prod' G_i$; let φ be a n.p.d.f. on G such that, denoting by φ_i its restriction to G_i, for every $g = (g_i) \in G$ the infinite product $\prod \varphi_i(g_i)$ is convergent, non zero and equal to $\varphi(g)$. Then there exists a unique isomorphism $\bigotimes_{i\in I}^{h(\xi_{\varphi_i})} H_{\varphi_i} \longrightarrow H$ carrying $\otimes \Lambda_{\varphi_i}(\delta_{g_i})$ into $\Lambda_\varphi(\delta_g)$ for every $g = (g_i) \in \prod' G_i$.

For $g = (g_i)$ and $h = (h_i)$ in $\prod' G_i$ we have

$$(\otimes \Lambda_{\varphi_i}(\delta_{g_i}) \mid \otimes \Lambda_{\varphi_i}(\delta_{h_i})) = \prod (\Lambda_{\varphi_i}(\delta_{g_i}) \mid \Lambda_{\varphi_i}(\delta_{h_i}))$$

$$= \prod \varphi_i(h_i^{-1} g_i) = \varphi(h^{-1}g)$$

$$= (\Lambda_\varphi(\delta_g) \mid \Lambda_\varphi(\delta_h))$$

thus we get an isomorphism of the tensor product of the H_{φ_i}'s onto the subspace of H generated by vectors of the form $\Lambda_\varphi(\delta_g)$, $g \in \prod' G_i$; it remains to prove that these vectors are total, or that every vector $\Lambda_\varphi(\delta_h)$, $h \in G$, can be approximated by such a vector. We define g by

$$g_i = \begin{cases} h_i & \text{if } i \in J \\ e_i & \text{if } i \notin J \end{cases}$$

where J is an arbitrary finite subset of I ; then

$$\| \Lambda_\varphi(\delta_g) - \Lambda_\varphi(\delta_h) \|^2 = 2 - 2 \operatorname{Re} \varphi(h^{-1}g)$$

$$= 2 - 2 \operatorname{Re} \prod_{i \notin J} \varphi_i(h_i)$$

and since $\prod_{i\in I} \varphi_i(h_i)$ is convergent and non zero, $\prod_{i\notin J} \varphi_i(h_i)$ tends to 1 when J tend to I, so that $\Lambda_\varphi(\delta_g) - \Lambda_\varphi(\delta_h) \longrightarrow 0$.

§ 3.2. Positive definite functions of type (S). Definition and first properties

Definition 3.3. We call representation of type (S) of a topological group G a (continuous unitary) representation U of G in a Hilbert space of the form SH such that $U(g) \in \mathfrak{S}_H$ $\forall g \in G$; the function $g \longmapsto (U(g) \text{ EXP } 0 \mid \text{EXP } 0)$ will be called continuous positive definite function of type (S) associated with U.

Warning : we do not suppose that EXP 0 is cyclic for U (see remark 4.2). The following is an attempt to determine all c.p.d.f. of type (S) on a given group G.

By theorem 2.1 to give U is equivalent to giving 3 continuous mappings

$$A : g \longmapsto A_g \in \mathcal{U}(H)$$
$$b : g \longmapsto b_g \in H$$
$$c : g \longmapsto c_g \in \underline{U}$$

satisfying the following relations

$$A_{gg'} = A_g A_{g'} \tag{3.1}$$
$$b_{gg'} = b_g + A_g b_{g'} \tag{3.2}$$
$$c_{gg'} = c_g c_{g'} \exp(i \operatorname{Im}(b_g \mid A_g b_{g'})) . \tag{3.3}$$

Relation (3.1) means that A is a representation of G in H ; (3.2) that b is a (continuous) 1-cocycle for A ; and (3.3) that the mapping $(g,g') \longmapsto \exp(i\ \mathrm{Im}(b_g|A_g b_{g'}))$ is the coboundary of the mapping c (G acting trivially in $\underline{U}$). The associated c.p.d.f. of type (S) is given by

$$\varphi(g) = c_g \exp(-\|b_g\|^2/2) \tag{3.4}$$

(For the definition of cocycles and coboundaries see App.B.)
Let us look more closely at condition (3.3) ; given a representation A and a 1-cocycle b we set

$$\alpha_{g,g'} = \exp(i\ \mathrm{Im}(b_g|A_g b_{g'})) ; \tag{3.5}$$

it is easy to check that α is a 2-cocycle, i.e. an element of $Z^2(G,\underline{U})$; and (3.3) means that

$$\partial c = \alpha . \tag{3.5'}$$

Such a mapping c, if it exists at all, is not unique : it is determined up to a character (continuous morphism of G into $\underline{U}$) ; so that the corresponding φ is also determined up to a character. Concerning the existence of c satisfying (3.5') we have the following results:

Proposition 3.1. If G is a simply-connected semi-simple Lie group, (3.5') has a solution for every representation A and cocycle b.

In fact each 2-cocycle in $Z^2(G,\underline{U})$ is a coboundary since the group $H^2(G,\underline{U})$ is trivial (see [44], ch.X,§5).

Proposition 3.2. If $b = \partial\omega$ where $\omega \in H$, the real function $(g,g') \longmapsto \mathrm{Im}(b_g|A_g b_{g'})$ is the coboudary of the real function $g \longmapsto \mathrm{Im}(A_g\omega|\omega) = \mathrm{Im}(b_g|\omega)$; therefore (3.5') has a solution, namely $c_g = \exp(i \operatorname{Im}(A_g\omega|\omega))$; the corresponding φ is given by $\varphi(g) = \exp(((A_g-I)\omega|\omega))$.

In fact we have

$$\begin{aligned}\operatorname{Im}(b_g|A_g b_{g'}) &= \operatorname{Im}(A_g\omega - \omega \mid A_gA_{g'}\omega - A_g\omega)\\ &= \operatorname{Im}((\omega|A_{g'}\omega) + (\omega|A_g\omega) - (\omega|A_{gg'}\omega))\\ &= \operatorname{Im}((A_{gg'}\omega|\omega) - (A_g\omega|\omega) - (A_{g'}\omega|\omega))\end{aligned}$$

(3.6).

Finally

$$\begin{aligned}\varphi(g) &= \exp(i \operatorname{Im}(A_g\omega|\omega) - \|A_g\omega - \omega\|^2/2)\\ &= \exp(-\|\omega\|^2 + i\operatorname{Im}(A_g\omega|\omega) + \operatorname{Re}(A_g\omega|\omega))\\ &= \exp(((A_g-I)\omega|\omega)).\end{aligned}$$

(For further results of this type see [1]).

<u>Corollary</u> 3.1. For every c.p.d.f. ψ on G, the function $\varphi(g) = \exp(\psi(g) - \psi(e))$ is a c.p.d.f. of type (S).

In fact ψ is of the form $\psi(g) = (A_g\omega|\omega)$.

<u>Theorem</u> 3.1. If G is compact, the continuous positive definite functions of type (S) on G are exactly the functions of the form $g \longmapsto \chi(g)\exp(\psi(g) - \psi(e))$ where χ is a character and ψ a continuous positive definite function.

In fact every 1-cocycle b is a coboundary (see App.B.)

<u>Remark</u> 3.2. Proposition 3.1 could be used to determine all c.p.d.f. of type (S) on a simply-connected semi-simple Lie group ; but unfortunately the study of the set Z^1 for these groups is far from

being achieved ! On the contrary this study is more or less achieved for nilpotent and solvable groups (see [1] and [15]), so that it seems likely that the results of the next paragraph, concerning commutative groups, can be generalized to nilpotent and solvable groups ; one could begin with the nilpotent group, the elements of which are triples $(u,v,w,) \in \underline{R}^3$ with the composition law

$$(u,v,w)\,(u',v',w') = (u+u',v+v',w+w'+uv') \;;$$

for this group the result is known only in the case where $\pi(0,0,w)$ is a scalar operator of the form e^{ikw} (we have set $\pi(g) = U_{A_g, b_g, c_g}$) and it can be obtained by the same method as th.3.2 (see [1]).

§ 3.3. The case of commutative locally compact groups

We suppose in this paragraph that G is a separable commutative locally compact group ; we consider a representation A of G in a separable Hilbert space H and we try to determine all the objects b,c satisfying (3.2) and (3.3) as well as the corresponding c.p.d. f. of type (S) given by (3.4).

Case 1. We suppose that A does not contain the trivial representation.

As in App.B.2 we write

$$H = \int_{\hat{G}}^{\oplus} H_\chi \, d\mu(\chi) \qquad (3.6')$$

so that A_g is the multiplication operator by the function $\chi \longmapsto \langle \chi ,g \rangle$; the 1-cocycles b correspond bijectively to the elements ω of $\bar{H}$ by the following relation : b_g has coordinates

$b_{g,\chi} = (<\chi,g>-1)\omega_\chi$. We fix ω in H and we set

$$d\nu(\chi) = \|\omega_\chi\|^2.d\mu(\chi) \quad ;$$

we recall that ν is a Radon measure on $\hat{G} - \{\varepsilon\}$ (ε = neutral element of $\hat{G}$) which is bounded on the complementary of every neighbourhood of ε . The corresponding α (see (3.5)) is given by

$$\alpha_{g,g'} = \exp\left[i \text{ Im}\int((<\chi,g>-1)\omega_\chi|<\chi,g>(<\chi,g'>-1)\omega_\chi).d\mu(\chi)\right]$$

$$= \exp\left[i \text{ Im}\int(<\chi,gg'> - <\chi,g> - <\chi,g'>).d\nu(\chi)\right]$$

which is rather analogous to the right hand side of (3.6) except for the fact that our ω might not belong to H ; we would like to set, as in prop.3.2,

$$c_g = \exp(i \text{ Im}(b_g|\omega))$$

$$= \exp\left[i \text{ Im}\int(b_{g,\chi}|\omega_\chi).d\mu(\chi)\right]$$

$$= \exp\left[i \text{ Im}\int(<\chi,g>-1).d\nu(\chi)\right]$$

but $(b_g|\omega)$ does not make sense in general and the function $<\chi,g>-1$ is not necessarily ν-integrable. To make it integrable we modify it by using the following result ([32], lemma 5.3).

Lemma 3.2. There exists a real continuous function J on $G\times\hat{G}$ with the following properties :

(i) $J(gg',\chi) = J(g,\chi) + J(g',\chi)$

(ii) for every compact $C\subset G$ there exists a neighbourhood V of ε such that for $g\in C$ and $\chi\in V$, $i\,J(g,\chi)$ is equal to the principal determination of $\log<\chi,g>$; we can then write

$$i\,J(g,\chi) = <\chi,g>-1 - \tfrac{1}{2}(<\chi,g>-1)^2(1 + h(g,\chi)) \qquad (3.7)$$

where h is a bounded function

(iii) for every compact $C \subset G$ the numbers $J(g,\chi)$ with $g \in C$ and $\chi \in \hat{G}$ are bouded.

For instance if $G = \underline{R}^n$ we can write $g = (g_1, \ldots g_n)$ and $\chi = (\chi_1, \ldots \chi_n)$ and take

$$J(g,\chi) = \sum_j g_j \chi_j J'(\chi)$$

where

$$J'(\chi) = \begin{cases} 1 & \text{if } \|\chi\| \le 1 \\ \|\chi\|^{-2} & \text{if } \|\chi\| > 1 \end{cases} ;$$

if $\hat{G}$ is totally disconnected (or equivalently if G is a union of compact subgroups) J is identically 0 because of (i).

The function $\chi \longmapsto \langle \chi, g\rangle - 1 - i\, J(g,\chi)$ is ν-integrable ; in fact let V a neighbourhood of ε such that (3.7) holds with $C = \{g\}$; our function is ν-integrable on $\hat{G} - V$ since on this set it is bouded as well as ν ; on V we have

$$\langle \chi, g\rangle - 1 - i\, J(g,\chi) = \tfrac{1}{2} (\langle \chi, g\rangle - 1)^2 (1 - h(g,\chi)) \qquad (3.8)$$

which is integrable since $(\langle \chi, g\rangle - 1)^2$ is integrable and $(1+h)$ is bounded. Therefore we can set

$$c_g = \exp\left[i \operatorname{Im} \int (\langle \chi, g\rangle - 1 - i\, J(g,\chi))\, d\nu(\chi) \right] \qquad (3.9)$$

and we have to show

a) that $\partial c = \alpha$ which is an easy computation by using property (i) in the lemma ; the proof actually shows that the real function $\operatorname{Im}(b_g | A_g b_{g'})$ is the coboundary of the real function $\operatorname{Im} \int (\langle \chi, g\rangle - 1 - i\, J(g,\chi)).d\nu(\chi)$.

b) that c is continuous, or, equivalently because of a), that c_g tends to 1 when g tends to e. We can suppose that g remains in a compact C ; choose V compact such that (3.7) holds ; take an $\eta > 0$. There exists a compact $K \supset V$ such that $\nu(\hat{G} - K) \leq \eta$; using (3.8) we have

$$\left|\int(<\chi,g> - 1 - i\,J(g,\chi)).d\nu(\chi)\right| \leq \left|\int_V \tfrac{1}{2}(<\chi,g> - 1)^2.\right.$$

$$\left.(1+h(g,\chi)).d\nu(\chi)\right| + \left|\int_{K-V}\right| + \left|\int_{\hat{G}-K}(<\chi,g> -1-iJ(g,\chi)d\nu\right| ;$$

the first integral tends to 0 since h is bounded and the integral $\int_{\hat{G}} |<\chi,g>-1|^2 \, d\nu(\chi)$ tends to 0 ; the second one tends to 0 too since on K-V, ν is bounded and $<\chi,g>-1$ as well as $J(g,\chi)$ converge uniformly to 0 ; the third one is less than η multiplied by a constant.

We have thus proved :

<u>Lemma</u> 3.3. Suppose that A does not contain the trivial representation ; then for every 1-cocycle b or equivalently for every ω in $\bar{H}$, the function $(g,g') \longmapsto \mathrm{Im}(b_g|A_g b_{g'})$ is the coboudary of the function $g \longmapsto \mathrm{Im}\int(<\chi,g> - 1 - i\,J(g,\chi)).d\nu(\chi)$; therefore (3.5') has a solution, given by (3.9) ; the corresponding φ is given by

$$\varphi(g) = \exp\left[\int(<\chi,g> - 1-i\,J(g,\chi)).d\nu(\chi)\right] \quad (3.10)$$

Conversely if ν is a positive measure on $\hat{G} - \{\varepsilon\}$ such that $\int|<\chi,g> - 1|^2.d\nu(\chi)$ is finite for every g and tends to 0 when g tends to e, the function φ defined by (3.10) is a c.p.d.f.

of type (S) : one has only to take for μ a normalized measure equivalent to ν , say $\mu = \gamma\nu$ where γ is a strictly positive function such that $\nu(\gamma) = 1$, and to set $H = L^2(G,\mu)$, $\omega = \gamma^{-\frac{1}{2}}$, A_g = multiplication operator by the function $\chi \longmapsto \langle\chi,g\rangle$; b_g = function $\chi \longmapsto (\langle\chi,g\rangle-1)\,\gamma(\chi)^{-\frac{1}{2}}$; c_g as in (3.9).

Case 2. We now suppose that A is trivial.

In that case a 1-cocycle b is nothing but a continuous morphism of G into H and the corresponding α is

$$\alpha_{g,g'} = \exp(i\ \mathrm{Im}(b_g | b_{g'})) ;$$

if (3.5') has a solution c we must have

$$\exp(i\ \mathrm{Im}(b_g | b_{g'})) = c_{gg'}/c_g\, c_{g'} = c_{g'g}/c_{g'}\, c_g$$

$$= \exp(i\ \mathrm{Im}(b_{g'} | b_g)) = \exp(-i\ \mathrm{Im}(b_g | b_{g'}))$$

whence

$$\mathrm{Im}(b_g | b_{g'}) \in \pi\,\underline{Z} \qquad \forall\ g,g' \in G . \tag{3.11}$$

We can give further information only in three particular cases :

a) if G is a union of compact subgroups, b is necessarily null, c is a character and φ is a character too.

b) if the function $\mathrm{Im}(b_g | b_{g'})$ is the coboundary of some function it is simultaneously symmetric and antisymmetric, hence null ; then the solutions of (3.5') are exactly the characters of G ; the corresponding φ is $\varphi(g) = \chi(g)\exp(-\|b_g\|^2/2)$ where χ is a character of G and b can be considered as an arbitrary continuous morphism of G into a real Hilbert space.

c) if G is connected (3.11) implies $\mathrm{Im}(b_g|b_{g'}) = 0$ and we have the same conclusions as in b).

<u>Definition</u> 3.4. Given a topological group G we call <u>continuous positive quadratic form</u> on G every function of the form $Q(g) = \|u(g)\|^2$ where u is a continuous morphism of G into a real Hilbert space. In the case of $G = \underline{R}^n$ we get the usual positive quadratic forms.

We have thus proved the following

<u>Lemma</u> 3.4. Suppose that A is trivial ; let b be a continuous morphism of G into H ; then

(i) if G is a union of compact subgroups, b is null, the solutions of (3.5') are exactly the characters of G as well as the corresponding φ

(ii) if the function $\mathrm{Im}(b_g|b_{g'})$ is the coboundary of some function, the solutions of (3.5') are exactly the characters of G and the corresponding φ are the functions $g \longmapsto \chi(g)\, e^{-Q(g)}$ where χ is a character of G and Q a continuous positive quadratic form on G

(iii) if G is connected, (3.5') has a solution if and only if $\mathrm{Im}(b_g|b_{g'}) = 0 \quad \forall\, g,g'$ and the conclusions of (ii) hold.

<u>Case</u> 3. General case.

We write $H = H' \oplus H''$ where H" is the set of all invariant vectors for A and $H' = H \ominus H''$; every 1-cocycle b can be written $b_g = b'_g + b''_g$ with $b'_g \in H'$, $b''_g \in H''$; moreover b' is a 1-cocycle for the representation in H' which does not contain the trivial one, and b" is a continuous morphism of G into H" ; the corresponding α can be decomposed into $\alpha = \alpha' \alpha''$ with

$$\alpha'_{g,g'} = \exp(i\ \mathrm{Im}(b'_g | A_g b'_{g'}))$$

$$\alpha''_{g,g'} = \exp(i\ \mathrm{Im}(b''_g | b''_{g'})) .$$

By lemma 3.2 there exists a function c' satisfying $\partial c' = \alpha'$; if the equation $\partial c = \alpha$ has a solution, the equation $\partial c'' = \alpha''$ has a solution too, namely $c'' = c/c'$; then $c = c'c''$ and the corresponding φ is the product of the corresponding φ' and φ''. We have thus proved the following result

Theorem 3.2. Let G be a separable commutative locally compact group which is either connected or a union of compact subgroups ; the continuous positive definite functions of type (S) on G which correspond to separable representations of type (S) are exactly the functions

$$\varphi(g) = \chi_0(g).\exp\left[-Q(g) - \int_{\hat{G}} (<\chi, g> - 1 - i\ J(g,\chi)).d\nu(\chi)\right] \tag{3.12}$$

where χ_0 is a character of G ; Q is a continuous positive quadratic form (necessarily null if G is a union of compact subgroups) ; ν is a positive Radon measure on $\hat{G} - \{\varepsilon\}$ such that the integral $\int |<\chi,g>-1|^2.d\nu(\chi)$ is finite for every $g \in G$ and tends to 0 when g tends to e ; and finally J is defined as in lemma 3.2 and is null if G is a union of compact subgroups. Moreover such a measure ν is bounded on the complementary of every neighbourhood of ε.

<u>Example</u> 3.2. For $G = \underline{R}^n$ the function J can be replaced by the function $\chi \longmapsto g.\chi/(1-\chi^2)$ where $g.\chi = \sum_j g_j \chi_j$ and $\chi^2 = \sum \chi_j^2$, which has the same effect. On the other hand ν can be written as $d\nu(\chi) = (1+\chi^2)/\chi^2 . d\sigma(\chi)$ where σ is an arbitrary finite positive measure without mass at 0 ; then (3.12) can be rewritten as

$$\varphi(g) = \exp\left[i\, g.\chi_0 - Q(g) + \int_{\underline{R}^n} (e^{i\, g.\chi} - 1 - i\, g.\chi/(1+\chi^2)).\frac{1+\chi^2}{\chi^2}\, d\sigma(\chi)\right]$$

where χ_0 is an element of $\underline{R}^n$, Q a positive quadratic form and σ a finite positive measure without mass at 0. We thus see that our p.d.f. is the product of 3 p.d.f.: a character, a Gaussian p.d.f. and a Poissonian p.d.f.(see App.E).

<u>Example</u> 3.3. For $G = \underline{Z}$, theorem 3.2 does not apply but we can proceed as follows : for the trivial part of the representation A, b_g can be written as $b_g = g\xi$ where $\xi \in H$; then $(b_g | b_{g'}) = gg' \|\xi\|^2$ is real, α is equal to 1, c is a character χ_0 and $\varphi(g) = \chi_0(g) \exp(-kg^2)$ where k is an arbitrary positive number. For the non-trivial part we can take $J(g,\chi) = g.F(\chi)$ where F is as in the figure. Finally we have a formula quite analogous to (3.12) :

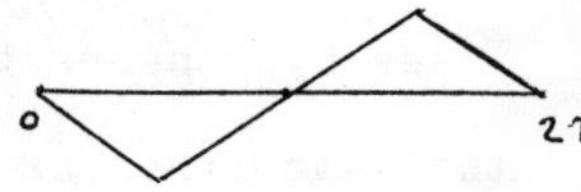

$$\varphi(g) = \exp\left[i\, g\chi_c - k\, g^2 + \int_{\underline{T}} (e^{ig\chi} - 1 - i\, g\, F(\chi)).d\nu(\chi)\right].$$

<u>Example</u> 3.4. For $G = \underline{Z}^2$ there exist c.p.d.f. of type (S) of a different kind, for instance (setting $g = (m,n)$)

$$\varphi(m,n) = \exp\left[i\pi\, m\, n - \pi^2\, m^2/2 - n^2/2\right]$$

which is obtained with $H = \underline{C}$, A trivial, $b_{m,n} = i\pi m + n$, $c_{m,n} = e^{i\pi mn}$.

<u>Remark</u> 3.3. Let us look more closely at the continuous positive quadratic forms on a group G (see def.3.4). By choosing an orthonormal basis ξ_α of the corresponding real Hilbert space we can write

$$u(g) = \sum_\alpha \theta_\alpha(g).\xi_\alpha$$

$$\| u(g)\|^2 = \sum_\alpha \theta_\alpha(g)^2$$

where the θ_α 's are real characters of G (i.e. continuous morphisms of G into $\underline{R}$) ; we are thus led to look at the set (vector space) of all real characters of G. It is proved in [34] § 4 that this vector space is finite dimensional if G is compactly generated ; and that in the general case it admits a countable basis (θ_n) such that for every compact $K \subset G$ almost all θ_n are identically zero on K.

<u>Remark</u> 3.4. In theorem 3.2 we have supposed G and H separable in order to use the desintegration (3.6'), i.e. Stone's theorem in its infinitesimal form ; but it is highly likely that one can drop these hypotheses by using the global form of Stone's theorem along the same lines as in the proof of th.2 in [15].

Chapter 4. CONDITIONALLY POSITIVE DEFINITE FUNCTIONS AND INFINITELY DIVISIBLE POSITIVE DEFINITE FUNCTIONS

§ 4.1. Conditionally positive definite matrices and kernels

Definition 4.1. A complex matrix with coefficients $a_{i,j}$, $i,j = 1,\dots n$, is said to be conditionally positive definite if (3.0) holds for any $x_1,\dots x_n$ satisfying $x_1 + \dots + x_n = 0$.

For $n = 2$ this means merely that $a_{11} + a_{22} - a_{12} - a_{21} \geq 0$; thus we see that a conditionally positive definite matrix is not necessarily hermitian.

Proposition 4.1. The matrix $(a_{i,j})$ is conditionally positive definite and hermitian iff the matrix $(e^{t\, a_{i,j}})$ is positive definite for any positive number t.

(See [12])

Proof of the necessity. We first prove that the quadratic form $\sum a_{ij} y_i \bar{y}_j$ is real and bounded below on the hyperplane $\sum y_i = 1$. In fact setting $x_1 = y_1 - 1$, $x_2 = y_2$,... $x_n = y_n$ we have $\sum x_i = 0$ and

$$\sum_{i,j=1}^{n} a_{ij} y_i \bar{y}_j = \sum_{i,j=2}^{n} a_{ij} x_i \bar{x}_j + \sum_{j=2}^{n} a_{1j}(x_1+1)\bar{x}_j$$

$$+ \sum_{i=2}^{n} a_{i1} x_i (\bar{x}_1+1) + a_{11}(x_1+1)(\bar{x}_1+1)$$

$$= \sum_{i,j=1}^{n} a_{ij} x_i \bar{x}_j + 2 \operatorname{Re} \sum_{i=2}^{n} a_{i1} x_i + a_{11}(2 \operatorname{Re} x_1 + 1)$$

which is real since our matrix is hermitian ; when x tends to infinity the sign is that of the first term, which is positive ; thus the expression is bounded below.

We can thus write $\sum a_{ij} x_i \bar{x}_j \geq K$ for $\sum x_i = 1$; then

$$\sum a_{ij} x_i \bar{x}_j \geq K |\sum x_i|^2 \qquad \forall \; x_1, \ldots x_n$$

$$\sum (1 + t\, a_{ij})\, x_i \bar{x}_j \geq |\sum x_i|^2 + t\, K |\sum x_i|^2$$

$$= (1 + t\, K) |\sum x_i|^2$$

which shows that for sufficiently small t the matrix $(1 + t a_{ij})$ is positive definite ; for arbitrary t and sufficiently large n the matrix $(1 + t\, a_{ij}/n)$ is positive definite ; so is the matrix $((1 + t\, a_{ij}/n)^n)$ and its limit $(e^{t\, a_{ij}})$.

<u>Proof of the sufficiency</u>. The matrix $(e^{t\, a_{ij}})$ being hermitian we have

$$e^{t\, a_{ij}} = \overline{e^{t\, a_{ji}}} = e^{t\, \overline{a_{ji}}}$$

for every t, whence $a_{ij} = \overline{a_{ji}}$. Now if $\sum x_i = 0$ the function $t \longmapsto \sum e^{t\, a_{ij}} x_i \bar{x}_j$ is positive for $t > 0$ and null for $t = 0$; its derivative at 0, equal to $\sum a_{ij} x_i \bar{x}_j$, must be positive.

<u>Definition</u> 4.2. A kernel K on a set E is <u>conditionally positive definite</u> if for any elements $u_1, \ldots u_n$ in E the matrix with coefficients $K(u_i, u_j)$ is conditionally positive definite.

<u>Proposition</u> 4.2. Let u_o be an arbitrary element of E ; for a kernel K the following conditions are equivalent :

(i) K is conditionally positive definite and $K(u_o, u_o) \leq 0$

(ii) the kernel $\tilde{K}(u,v) = K(u,v) - K(u,u_o) - K(u_o,v)$ is positive definite.

(See [22])

<u>Proof of</u> (i) $\Longrightarrow$ (ii). Take $u_1, \dots u_n \in E$ and $x_1, \dots x_n \in \underline{C}$; set $x_o = -x_1 - \dots -x_n$; we have

$$0 \leq \sum_{i,j=o}^{n} K(u_i,u_j)\, x_i \bar{x}_j$$

$$= \sum_{i,j=1}^{n} K(u_i,u_j)\, x_i \bar{x}_j - \sum_{j=1}^{n} K(u_o,u_j) \sum_{i=1}^{n} x_i \bar{x}_j$$

$$- \sum_{i=1}^{n} K(u_i,u_o) \sum_{j=1}^{n} x_i \bar{x}_j + K(u_o,u_o)\, x_o \bar{x}_o$$

$$= \sum_{i,j=1} \left[K(u_i,u_j) - K(u_o,u_j) - K(u_i,u_o)\right] x_i \bar{x}_j + K(u_o,u_o) x_o \bar{x}_o$$

and the assertion follows since $K(u_o,u_o) \leq 0$.

<u>Proof of</u> (ii) $\Longrightarrow$ (i). We have $K(u_o,u_o) \leq 0$ since

$$0 \leq \tilde{K}(u_o,u_o) = - K(u_o,u_o) ;$$

moreover for $u_1, \dots u_n \in E$, $x_1, \dots x_n \in \underline{C}$, $\sum x_i = 0$ we have

$$\sum K(u_i,u_j)\, x_i \bar{x}_j = \sum \tilde{K}(u_i,u_j)\, x_i \bar{x}_j + \sum K(u_i,u_o)\, x_i \bar{x}_j +$$

$$\sum K(u_o,u_j)\, x_i \bar{x}_j$$

$$= \sum \tilde{K}(u_i,u_j)\, x_i \bar{x}_j \geq 0 .$$

§ 4.2. Conditionally positive definite functions on groups

Definition 4.3. A complex function ψ on a group G is conditionally positive definite if the kernel $K(g,g') = \psi(g'^{-1}g)$ is conditionally positive definite. It is hermitian if $\psi(g^{-1}) = \overline{\psi(g)}$.

Proposition 4.3. A function ψ on G is conditionally positive definite and hermitian iff the function $e^{t\psi}$ is positive definite for every positive number t.

Follows directly from prop.4.1.

Proposition 4.4. A function ψ on G is conditionally positive definite and negative at e (neutral element of G) iff the kernel $(g,g') \longmapsto \psi(g'^{-1}g) - \psi(g) - \psi(g'^{-1})$ is positive definite.

Follows from prop.4.2.

Our next aim is to determine the general form of the continuous hermitian conditionally positive definite functions (c.h.c.p.d.f.) null at e on a topological group G.

Proposition 4.5. Let G be a topological group and ψ a c.h.c.p.d.f. null at e ; there exist a Hilbert space H and a triple (A,b,c) satisfying (3.1),(3.2),(3.3) as well as the following relations :

$$(b_g | b_{g'}) = \psi(g'^{-1}g) - \psi(g) - \psi(g'^{-1}) \tag{4.1}$$

$$c_g \exp(-\|b_g\|^2/2) = e^{\psi(g)}. \tag{4.2}$$

In particular the positive definite function e^{ψ} is of type (S). Moreover the real function $(g,g') \longmapsto \mathrm{Im}(b_g | A_g b_{g'})$ is the coboundary of the function $\mathrm{Im}\,\psi$.

(See [1] th.5.1 and [34])

As in the Gelfand-Segal construction (§ 3.1) we form $\underline{C}^{(G)}$ and we endow it with a scalar product

$$(f|f') = \sum_{g,g'} (\psi(g'^{-1}g) - \psi(g) - \psi(g'^{-1})).f(g).\overline{f'(g')} ;$$

this sesquilinear form is actually hermitian since ψ is so, and positive due to prop.4.4 ; in particular

$$(\delta_g | \delta_{g'}) = \psi(g'^{-1}g) - \psi(g) - \psi(g'^{-1}) \qquad (4.3)$$

$$(\delta_g | \delta_g) = - 2 \operatorname{Re} \psi(g) .$$

We define for every h in G a linear operator T_h in $\underline{C}^{(G)}$ by

$$T_h \delta_g = \delta_{hg} - \delta_h ;$$

T_h preserves the scalar product since

$$(T_h \delta_g | T_h \delta_{g'}) = (\delta_{hg} - \delta_h | \delta_{hg'} - \delta_h) = (\delta_g | \delta_{g'}) ;$$

moreover we have $T_{hh'} = T_h T_{h'}$. Let H be the Hilbert space obtained by separation and completion of $\underline{C}^{(G)}$, Λ the canonical mapping $\underline{C}^{(G)} \longrightarrow H$, A_h the unitary operator in H corresponding to T_h :

$$A_h(\Lambda(\delta_g)) = \Lambda(\delta_{hg}) - \Lambda(\delta_h) ;$$

then A is a unitary representation of G in H, which is continuous since if h tends to e we have

$$\| A_h(\Lambda(\delta_g)) - \Lambda(\delta_g)\|^2 = 2 \|\Lambda(\delta_g)\|^2 - 2 \operatorname{Re}(A_h(\Lambda(\delta_g)) | \Lambda(\delta_g))$$

$$= - 4 \operatorname{Re} \psi(g) - 2 \operatorname{Re} [\psi(g^{-1}hg) - \psi(hg) - \psi(g^{-1}h) + \psi(h)$$

$$\longrightarrow - 4 \operatorname{Re} \psi(g) - 2 \operatorname{Re} [- \psi(g) - \psi(g^{-1})] = 0.$$

We now set $b_g = \Lambda(\delta_g)$; b is a cocycle since

$$b_{gg'} = \Lambda(\delta_{gg'}) = A_g(\Lambda(\delta_{g'})) + \Lambda(\delta_g) = A_g b_{g'} + b_g \;;$$

it is continuous since if g tends to e

$$\|b_g\|^2 = -2 \operatorname{Re} \psi(g) \longrightarrow 0 .$$

Finally we set

$$c_g = \exp(\tfrac{1}{2}(\psi(g) - \psi(g^{-1}))) = \exp(i \operatorname{Im} \psi(g)) \;;$$

c is a continuous mapping of G into $\underline{U}$ and satisfies (3.3) as easily verified ; (4.1) follows from (4.3) and (4.2) is trivial.

Proof of the last assertion :

$$\begin{aligned} \operatorname{Im}(b_g | A_g b_{g'}) &= \operatorname{Im}(A_g^{-1} b_g | b_{g'}) = -\operatorname{Im}(b_{g^{-1}} | b_{g'}) \\ &= -\operatorname{Im}(\psi(g'^{-1}g^{-1}) - \psi(g^{-1}) - \psi(g'^{-1})) \\ &= \operatorname{Im}(\psi(gg') - \psi(g) - \psi(g')). \end{aligned}$$

<u>Remark</u> 4.1. Prop.4.5 admits a partial converse : let H be a Hilbert space and (A,b,c) a triple satisfying (3.1),(3.2),(3.3) ; suppose that there exists a continuous real function γ such that

$$\operatorname{Im}(b_g | A_g b_{g'}) = \gamma(gg') - \gamma(g) - \gamma(g') \;;$$

define ψ by

$$\psi(g) = -\|b_g\|^2/2 + i\gamma(g) \;;$$

then it is easily verified that ψ is a c.h.c.p.d.f. null at e and satisfies (4.1) ; moreover $e^{i\gamma}/c$ is a character χ since

$$e^{i\gamma(gg')}/e^{i\gamma(g)}\, e^{i\gamma(g')} = e^{i \operatorname{Im}(b_g | A_g b_{g'})} = c_{gg'}/c_g c_{g'} \;;$$

then we have, instead of (4.2) : $\chi(g).c_g.\exp(-\|b_g\|^2/2) = e^{\psi(g)}$.

Remark 4.2. Let φ be the p.d.f. e^{ψ} and V the representation of G in SH defined by $V_g = U_{A_g, b_g, c_g}$; by (4.2) we have

$$\varphi(g) = (V_g \text{ EXP } 0 \mid \text{EXP } 0) ;$$

therefore U_φ can be embedded in V in such a manner that ξ_φ becomes EXP 0 ; but it must be noted that in general EXP 0 is <u>not cyclic</u> for V : if for instance G is finite and ψ is not additive, then (4.1) shows that $(b_g \mid b_{g'}) \neq 0$ for some g,g' ; thus His not reduced to 0, SH is infinite dimensional and V cannot have a cyclic vector.

The case of compact groups

Theorem 4.1. Let G be a compact group ; the continuous hermitian conditionally positive definite functions on G null at e are exactly the functions $g \longmapsto \varphi(g) - \varphi(e)$ where φ is a continuous positive definite function.

Let ψ be a c.h.c.p.d.f. null at e and use the notations of prop.4.5 ; by this prop. we know that $\text{Im}(b_g \mid A_g b_{g'})$ is the coboundary of $\text{Im}\,\psi(g)$; by App.B.2, b is the coboundary of a vector in H ; by prop.3.2, $\text{Im}(b_g \mid A_g b_{g'})$ is also the coboundary of the function $\text{Im}(A_g \omega \mid \omega)$; therefore $\text{Im}\,\psi(g) - \text{Im}(A_g \omega \mid \omega)$ is a continuous morphism of G into <u>R</u> ; it is null since G is compact. We then have

$$\begin{aligned} \psi(g) &= \text{Re}\,\psi(g) + i\,\text{Im}\,\psi(g) \\ &= -\|b_g\|^2/2 + i\,\text{Im}(A_g \omega \mid \omega) \\ &= ((A_g - I)\omega \mid \omega) = \varphi(g) - \varphi(e) \end{aligned}$$

where $\varphi(g) = (A_g \omega | \omega)$. Conversely for every c.p.d.f. φ, $\varphi(g) - \varphi(e)$ is a c.h.c.p.d.f. null at e (easy verification).

The case of commutative locally compact groups

Let G be a separable commutative locally compact group and ψ a c.h.c.p.d.f. null at e ; we use the notations of prop.4.5 ; H is separable since G is so and the b_g's are total in H. As in the proof of th.3.2 (case 3) we separate the trivial and non-trivial parts of H : $H = H' \oplus H''$, and we write $b_g = b'_g + b''_g$; we have

$$(b_g | A_g b_{g'}) = (b'_g | A_g b'_{g'}) + (b''_g | b''_{g'}) .$$

By prop.4.5 the function $\mathrm{Im}(b_g | A_g b_{g'})$ is the coboundary of $\mathrm{Im}\,\psi$; by lemma 3.3, $\mathrm{Im}(b'_g | A_g b'_{g'})$ is the coboundary of the function $\mathrm{Im}\int(<\chi, g> - 1 - i\,J(g,\chi)).d\nu(\chi)$; then $\mathrm{Im}(b''_g | b''_{g'})$ is the coboundary of the function

$$\gamma(g) = \mathrm{Im}\,\psi(g) - \mathrm{Im}\int(<\chi, g> - 1 - i\,J(g,\chi)).d\nu(\chi) ;$$

as already noticed in lemma 3.4, $\mathrm{Im}(b''_g | b''_{g'})$ is null ; therefore γ is a real character of G and we have

$$\begin{aligned}\psi(g) &= \mathrm{Re}\,\psi(g) + i\,\mathrm{Im}\,\psi(g) \\ &= -\|b_g\|^2/2 - i\left[\gamma(g) + \mathrm{Im}\int(<\chi, g> - 1 - i\,J(g,\chi))d\nu(\chi)\right] \\ &= i\,\gamma(g) - \|b''_g\|^2/2 + \int(<\chi, g> - 1 - i\,J(g,\chi)).d\nu(\chi).\end{aligned}$$

Theorem 4.2. Let G be a separable commutative locally compact group ; the continuous hermitian conditionally positive definite functions on G null at e are exactly the functions

$$\psi(g) = i\,\gamma(g) - Q(g) + \int_{\hat{G}} (<\chi, g> - 1 - i\,J(g,\chi)).d\nu(\chi) \qquad (4.4)$$

where χ is a real character of G and Q, ν ,J are as in th.3.2. (See [34], th.4.4.)

There remains only to be proved that for each χ , Q and ν , formula (4.4) defines a c.h.c.p.d.f. null at e, which is trivial.

Example 4.1. For $G = \underline{R}^n$ according to what has been said in example 3.2, (4.4) can be rewritten

$$\psi(g) = i\, g.\chi_o - Q(g) + \int_{\underline{R}^n} (e^{i\, g.\chi} - 1 - i\, g.\chi/(1+\chi^2)).\frac{1+\chi^2}{\chi^2}\, d\sigma(\chi) ;$$

this formula is sometimes referred to as "Levy-Khinchin formula".

Example 4.2. If G is a union of compact subgroups (for instance if G is a p-adic group $\underline{Q}_p$) (4.4) takes the simpler form

$$\psi(g) = \int_{\hat{G}} (<\chi, g> - 1).d\nu(\chi).$$

Remark 4.3. If the c.h.c.p.d.f. of th.4.2 is real, it is equal to its real part and formula (4.4) becomes

$$\psi(g) = - Q(g) + \int_{\hat{G}} \mathrm{Re}(<\chi, g> - 1).d\nu(\chi)$$
$$= - Q(g) - \int_{\hat{G}} |<\chi, g> - 1|^2.d\nu'(\chi)$$

where $\nu' = \nu/2$. This formula can be proved by a more direct method (see [20]) ; but one can hope that the present method can be extended to other groups than those which are considered here.

§ 4.3. Infinitely divisible positive definite functions

Definition 4.4. A continuous positive definite function φ on a topological group G is said to be infinitely divisible if for every positive integer n there exists a continuous positive definite function ω satisfying $\omega^n = \varphi$.

Example 4.3. If ψ is a continuous hermitian conditionally positive definite function, e^{ψ} is an i.d.c.p.d.f. (cf.prop.4.3). We shall prove later a partial converse of this result.

We now consider triples (H,U,ξ) where H is a Hilbert space, U a (continuous unitary) representation of G in H and ξ a unit cyclic vector for U ; we recall that two triples (H,U,ξ) and (H',U',ξ') are called equivalent if there exists an isomorphism of H onto H' carrying U into U' and ξ into ξ'.

Definition 4.5. A triple (H,U,ξ) is infinitely divisible is for every positive integer n there exists a triple (H_1,U_1,ξ_1) such that (H,U,ξ) is equivolent to $(K,U_1^{\otimes n}|K, \xi_1^{\otimes n})$ where K is the closed subspace of $H_1^{\otimes n}$ generated by the elements $U_1^{\otimes n}(g).\xi_1^{\otimes n}$ (It should be noted that for $n > 1$ K is distinct from $H_1^{\otimes n}$ since it is included in $S^n H_1$). Then a c.p.d.f. is infinitely divisible iff the associated triple does so.

Proposition 4.6. If φ is an infinitely divisible continuous positive definite function, the set of all g such that $\varphi(g) \neq 0$ is a (open and closed) subgroup.

(See [33], lemma 4.2.)

We can suppose $\varphi(e) = 1$; clearly $\varphi(g) \neq 0$ implies $\varphi(g^{-1}) = \overline{\varphi(g)} \neq 0$. Let us now prove that $\varphi(g_1) \neq 0$ and $\varphi(g_2) \neq 0$ imply $\varphi(g_1 g_2) \neq 0$. We first remark that if x is a positive number we have

$$\overline{\lim_{n=\infty}} \; n\,(1 - x^{1/n}) < \infty \iff x > 0 . \tag{4.5}$$

Moreover for every p.d.f. ω we have

$$(1 - \mathrm{Re}\,\omega(g_1 g_2))^{\frac{1}{2}} \leq (1 - \mathrm{Re}\,\omega(g_1))^{\frac{1}{2}} + (1 - \mathrm{Re}\,\omega(g_2))^{\frac{1}{2}} \; ; \tag{4.6}$$

in fact writing

$$\omega(g) = (U_g \xi \mid \xi)$$

we get

$$\| U_g \xi - \xi \| = (2 - 2\,\mathrm{Re}\,\omega(g))^{\frac{1}{2}}$$

$$\| U_{g_1 g_2} \xi - \xi \| \leq \| U_{g_1 g_2} \xi - U_{g_1} \xi \| + \| U_{g_1} \xi - \xi \|$$

$$= \| U_{g_2} \xi - \xi \| + \| U_{g_1} \xi - \xi \| .$$

Now let ψ be such that $\psi^n = \varphi$; ψ is positive definite and so are $\overline{\psi}$ and $|\psi|^2 = \psi\overline{\psi}$; applying (4.6) to the latter we obtain

$$(1 - |\psi(g_1 g_2)|^2)^{\frac{1}{2}} \leq (1 - |\psi(g_1)|^2)^{\frac{1}{2}} + (1 - |\psi(g_2)|^2)^{\frac{1}{2}}$$

$$(1 - |\varphi(g_1 g_2)|^{2/n})^{\frac{1}{2}} \leq (1 - |\varphi(g_1)|^{2/n})^{\frac{1}{2}} + (1 - |\varphi(g_2)|^{2/n})^{\frac{1}{2}}$$

then (4.5) proves that $(g_1 g_2) \neq 0$.

<u>Corollary</u> 4.1. If G is connected, a non identically vanishing infinitely divisible continuous positive definite function is nowhere vanishing.

Remark 4.4. If H is an open subgroup of G and φ a c.p.d.f. on H, the function $\tilde{\varphi}$ extending φ by 0 outside of H is again a c.p.d.f. In fact choose in each residue class gH a representative γ_a where a runs over some set A ; for any $g_1, \dots g_n \in G$ and $c_1, \dots c_n \in \underline{C}$ we have

$$\sum_{i,j=1}^{n} c_i \bar{c}_j \tilde{\varphi}(g_j^{-1} g_i) = \sum_a \sum_{g_i, g_j} c_i \bar{c}_j \varphi(g_j^{-1} g_i)$$

$$= \sum_a \sum_{g_i, g_j} c_i \bar{c}_j \varphi((\gamma_a^{-1} g_j)^{-1} (\gamma_a^{-1} g_i))$$

where g_i and g_j run over $\gamma_a H$; but each sum occuring under the sign $\sum_a$ is positive.

Proposition 4.7. If G is an arcwise connected group, every normalized infinitely divisible continuous positive definite function on G is the exponential of some continuous hermitian conditionally positive definite function null at e.

Let φ be such a p.d.f. and φ_n a c.p.d.f. such that $\varphi_n^n = \varphi$; since φ admits a continuous n-th root for every n, it admits also a continuous logarithm ψ and we can suppose $\psi(e) = 0$. To prove that ψ is hermitian and conditionally positive definite it is sufficient by prop.4.3 to prove that $e^{t\psi}$ is positive definite for every positive number t. We first take $t = 1/n$; then

$$(e^{\psi/n})^n = e^{\psi} = \varphi = \varphi_n^n$$

therefore $e^{\psi/n}/\varphi_n$ is a n-th root of 1, hence a constant k since G is connected ; moreover

$$k = e^{\psi(e)/n} / \varphi_n(e) = 1 ;$$

thus $e^{\psi/n} = \varphi_n$ is positive definite ; it follows that $e^{t\psi}$ is positive definite for every t of the form m/n , and by continuity for every t .

<u>Corollary</u> 4.2. Let G be an arcwise connected topological group.
(i) Every normalized infinitely divisible continuous positive definite function on G is of type (S).
(ii) If (H,U,ξ) is an infinitely divisible triple, U can be embedded in a representation V of type (S) in such a manner that ξ becomes EXP 0.

Follows directly from prop. 4.7 and 4.5. Note that in general U will be a strict subrepresentation of V (see remark 4.2).

Assertion (ii) is sometimes referred to as "Araki-Woods embedding theorem".

The following theorem is an immediate consequence of prop.4.7 and th.4.2 :

<u>Theorem</u> 4.3. Let G be an arcwise connected separable commutative locally compact group ; the normalized infinitely divisible continuous positive definite functions on G are exactly the functions

$$\varphi(g) = \exp\left[i\,\gamma(g) - Q(g) + \int_{\hat{G}} (\langle \chi, g \rangle - 1 - i\,J(g,\chi)).d\nu(\chi)\right] \tag{4.7}$$

where γ , Q, ν , J are as in th.4.1.
(See [32], th.7.1.)

<u>Corollary</u> 4.3. With the assumptions of the above theorem, every continuous positive definite function of type (S) corresponding to a separable representation of type (S) is the product of a character by a normalized infinitely divisible continuous positive definite function.

Note that a character is always a c.p.d.f. of type (S) but not always infinitely divisible ! If every character of G is infinitely divisible (one then says that $\hat{G}$ is divisible), then the two classes of p.d.f. are identical.

<u>Problem</u>. Can cor.4.3 be proved directly and generalized to other groups ?

<u>Example</u> 4.4. For $G = \underline{R}^n$ formula (4.7) is known as the <u>Lévy-Khinchin formula</u> (cf. e.g. [10], ch.III, § 4).

For the case of compact groups we have

<u>Theorem</u> 4.4. Let G be an arcwise connected compact group ; the normalized infinitely divisible continuous positive definite functions on G are exactly the functions $g \longmapsto e^{\psi(g) - \psi(e)}$ where ψ is a continuous positive definite function on G.

Follows from prop.4.7 and th.4.1.

There is a corollary quite analogous to corollary 4.3 ,

<u>Remark</u> 4.5. For a compact G, K.R.Parthasarathy has proved in [33] that the normalized infinitely divisible continuous positive definite functions are exactly the functions

$$g \longmapsto \begin{cases} \chi(g).e^{\psi(g)-\psi(e)} & \text{for } g \in H \\ 0 & \text{for } g \notin H \end{cases}$$

where χ is an infinitely divisible character of G, H an open subgroup of G and ψ a c.p.d.f. on H.

Remark 4.6. Remark 4.1 can be generalized as follows : let A be a representation of G in a space H and b a 1-cocycle for A ; we do not assume that the function $s(g,g') = \mathrm{Im}(b_g | A_g b_{g'})$ is a coboundary. The mapping $g \longmapsto U_{A_g, b_g, 1}$ is a projective representation V of G in SH with multiplier e^{is}. Set $\psi(g) = -\|b_g\|^2/2$; we have

$$(V_g \text{ EXP } 0 \mid \text{EXP } 0) = e^{\psi(g)}$$

$$(b_g | b_{g'}) = \psi(g'^{-1}g) - \psi(g) - \psi(g'^{-1}) + i\, s(g'^{-1},g) ;$$

it follows that ψ is hermitian, null at e and conditionally s-positive definite which means that the kernel

$$(g,g') \longmapsto \psi(g'^{-1}g) - \psi(g) - \psi(g'^{-1}) + i\, s(g'^{-1},g)$$

is positive definite. Conversely every such ψ can be obtained in this way ([34], th.2.1) ; this result can be used to generalize our th.4.2 to conditionally s-positive definite functions ([34], th.4.3) and our cor.4.2 to projective representations ([34], th. 5.4).

Chapter 5. BOOLEAN ALGEBRAS OF TENSOR DECOMPOSITIONS OF A HILBERT SPACE

This chapter is devoted to a deep theorem of Araki and Woods [2] about what we call Boolean algebras of tensor decompositions of Hilbert spaces ; this result will be used later in several circumstances.

Definition 5.1. Let H be a (complex) Hilbert space and ω a unit vector in H ; we call Boolean algebra of tensor decompositions (BATD) of the pair (H,ω) a family of four objects Θ , $(H_\theta)_{\theta\in\Theta}$, $(\omega_\theta)_{\theta\in\Theta}$, $(\Lambda_{\mathcal{F}})_{\mathcal{F}\in\tilde{\Theta}}$ where

- Θ is an abstract complete Boolean algebra (see App.C)
- for each $\theta\in\Theta$, H_θ is a Hilbert space and ω_θ a unit vector in H_θ
- $\tilde{\Theta}$ is the set of all partitions i.e. families $(\theta_i)_{i\in I}$ of pairwise disjoint elements of Θ
- for each $\mathcal{F} = (\theta_i)_{i\in I} \in \tilde{\Theta}$, $\Lambda_{\mathcal{F}}$ is an isomorphism of the tensor product of the spaces H_{θ_i} corresponding to the vectors ω_{θ_i} , onto $H_{\vee\theta_i}$ carrying $\otimes\,\omega_{\theta_i}$ into $\omega_{\vee\theta_i}$

satisfying the following conditions :

(i) $H_{\mathbb{1}} = H$, $\omega_{\mathbb{1}} = \omega$

(ii) $H_{\mathbb{O}} = \underline{C}$, $\omega_{\mathbb{O}} = 1$

(iii) (associativity condition) consider a partition $\mathcal{F} = (\theta_i)_{i\in I}$ of an element θ and for each i a set J_i and a partition $\mathcal{G}_i = (\theta_{i,j})_{j\in J_i}$ of θ_i ; set $J = \sum_{i\in I} J_i$ and $\mathcal{G} = (\theta_{i,j})_{(i,j)\in J}$ which is a partition of θ ; according to the associativity of the infinite tensor products (see App.A.2), we can write

$$\overset{h}{\underset{(i,j)\in J}{\otimes}}{}^{(\omega_{\theta_{i,j}})} H_{\theta_{i,j}} = \overset{h}{\underset{i\in I}{\otimes}}{}^{(\alpha_i)} \left(\overset{h}{\underset{j\in J_i}{\otimes}}{}^{(\omega_{\theta_{i,j}})} H_{\theta_{i,j}} \right)$$

where $\alpha_i = \underset{j\in J_i}{\otimes} \omega_{\theta_{i,j}}$; thus we can consider the isomorphism

$$\underset{i\in I}{\otimes} \Lambda_{\mathcal{G}_i} : \overset{h}{\underset{(i,j)\in J}{\otimes}}{}^{(\omega_{\theta_{i,j}})} H_{\theta_{i,j}} \longrightarrow \overset{h}{\underset{i\in I}{\otimes}}{}^{(\omega_{\theta_i})} H_{\theta_i} .$$

Then the required condition reads as follows :

$$\Lambda_{\mathcal{F}} \circ \underset{i\in I}{\otimes} \Lambda_{\mathcal{G}_i} = \Lambda_{\mathcal{G}} .$$

(iv) (commutativity condition) consider a partition $\mathcal{F} = (\theta_i)_{i\in I}$ of an element θ and a permutation s of the set I ; set $\theta'_i = \theta_{s(i)}$; by App.A.2 there exists a canonical isomorphism

$$A : \overset{h}{\underset{i\in I}{\otimes}}{}^{(\omega_{\theta_i})} H_{\theta_i} \longrightarrow \overset{h}{\underset{i\in I}{\otimes}}{}^{(\omega_{\theta'_i})} H_{\theta'_i} .$$

The required condition is $\Lambda_{\mathcal{F}'} \circ A = \Lambda_{\mathcal{F}}$.

Example 5.1. Consider Θ as above, a Hilbert space K and a mapping $\theta \longmapsto K_\theta$ where

- K_θ is a closed linear subspace of K
- $K_{\mathbb{1}} = K$, $K_{\mathbb{0}} = \{0\}$
- for every partition $\theta = \vee \theta_i$ the K_{θ_i} are mutually orthogonal and their sum is equal to K_θ

(such a mapping $\theta \longmapsto K_\theta$ will be called Boolean algebra of direct sum decompositions of K). Set $H = SK$ and $\omega = EXP\ 0$; then we get a BATD of (H,ω) by taking $H_\theta = S\,K_\theta$, $\omega_\theta = EXP\ 0$ and $\Lambda_{\mathcal{F}}$ equal to the canonical isomorphism $\otimes S\,K_{\theta_i} \longrightarrow S\,K_\theta$ described in prop.2.3.

The aim of the next theorem is to prove the converse, but of course under certain additional assumptions ; for instance Θ must be non atomic ; in fact we get a BATD by taking $H = \underline{C}^2$, ω arbitrary and Θ reduced to $\mathbb{0}$ and $\mathbb{1}$; but $\underline{C}^2$ is not the symmetric of any Hilbert space !

Definition 5.2. With the notations of definition 5.1 a non zero vector x in H will be called factorizable if for each finite partition of $\mathbb{1}$: $\mathcal{F} = (\theta_i)_{i\in I}$ there exist vectors x_i in H_{θ_i} such that $x = \Lambda_{\mathcal{F}}(\bigotimes_{i\in I} x_i)$. We shall denote by H^1 the set of all factorizable vectors and by H^0 the set of the factorizable vectors x satisfying $(x|\omega) = 1$; H^1 and H^0 are closed in H (see App.A).

Theorem 5.1. Consider a Boolean algebra of tensor decompositions as in definition 5.1 ; suppose that Θ is non atomic and that H^0

is total in H. Then there exist a Hilbert space K, a family of subspaces K_θ as in example 5.1 and a family of isomorphisms Φ_θ : $H_\theta \longrightarrow S K_\theta$ such that

(i) $\Phi_\theta(\omega_\theta) = \text{EXP } 0$

(ii) $\Phi_{1}(H^o) = \{\text{EXP } a \mid a \in K\}$

(iii) $\Phi_{1}(H^1) = \{\lambda \text{ EXP } a \mid \lambda \in \underline{C} , a \in K\}$

(iv) for every partition $\mathcal{F} = (\theta_i)_{i \in I}$, $\Phi_{\vee\theta_i} \circ \Lambda_{\mathcal{F}} \circ (\otimes \Phi_{\theta_i})^{-1}$ is equal to the canonical isomorphism $\overset{h\,(\text{EXP } 0)}{\underset{i \in I}{\otimes}} S K_{\theta_i} \longrightarrow SK_{\vee\theta_i}$.

(See [2], th.6.1) In th.8.1 we shall give an interpretation of this result in terms of continuous tensor products of Hilbert spaces.

<u>Sketch of the proof</u>

Suppose the problem is solved and write $H_\theta = S K_\theta$, $\omega_\theta =$ EXP 0 ; clearly every element of the form $x = \lambda$ EXP a is factorizable ; conversely it can be proved that every factorizable vector is of that form. Then among the factorizable vectors, those of the form EXP a are characterized by $(x \mid \omega) = 1$, so that EXP is a bijection of K onto H^o. Returning to our problem we see that K must be a set with a bijection $K \longrightarrow H^o$ which we shall denote by U. Take an element a in K ; as we shall see in the proof there exists for each θ a unique element x_θ in H_θ such that for each finite partition of 1 : $\mathcal{F} = (\theta_i)_{i \in I}$ we have $Ua = \Lambda_{\mathcal{F}}(\otimes x_{\theta_i})$ and moreover $(x_\theta \mid \omega_\theta) = 1$; in particular for a partition $\mathcal{F}$ of the form $\mathcal{F} = (\theta, \theta')$ we have $Ua = \Lambda_{\mathcal{F}}(x_\theta \otimes x_{\theta'})$; at the end of the proof x_θ will be EXP a_θ where a_θ is the projection of a into K_θ ; then we define K_θ as the set of all a such that $x_{\theta'} = \omega_{\theta'}$.

Finally we have to endow K with

1) a scalar product : $(a|b)$ must be a logarithm of $(Ua|Ub)$ but of course the difficulty lies in the choice of the log

2) a linear structure, i.e. to define $\alpha a + \beta b$ in terms of Ua and Ub. Take a "very fine" finite partition of $\mathbb{1}$: $\mathcal{F} = (\theta_i)_{i\in I}$; define $x_i = x_{\theta_i}$ as above and similarly y_i with respect to b ; supposing the problem solved and denoting by a_i and b_i the projections of a and b into K_{θ_i} , we have, since a_i and b_i are "very small"

$$x_i = \text{EXP } a_i \ \# \ \omega_i + a_i$$

$$y_i = \text{EXP } b_i \ \# \ \omega_i + b_i$$

$$\begin{aligned} \text{EXP } (\alpha a + \beta b) &= \otimes \text{EXP } (\alpha a_i + \beta b_i) \\ &\# \otimes (\omega_i + \alpha a_i + \beta b_i) \\ &\# \otimes (\omega_i + \alpha (x_i - \omega_i) + \beta (y_i - \omega_i)) \\ &= \otimes (\omega_i + \alpha x_i' + \beta y_i') \end{aligned}$$

where we have set $x_i' = x_i - \omega_i$, $y_i' = y_i - \omega_i$. Whence the idea of defining $\alpha a + \beta b$ by the relation

$$U(\alpha a + \beta b) = \lim_{\mathcal{F}} \otimes (\omega_i + \alpha x_i' + \beta y_i')$$

where $\mathcal{F}$ runs over the ordered set of all finite partitions of $\mathbb{1}$.

For the proof of th.5.1 we need several lemmas.

<u>Lemma</u> 5.1. Let x be an element of H^o ; there exist uniquely determined vectors $x_\theta \in H_\theta$ with the following properties :

(i) $(x_\theta | \omega_\theta) = 1$, $x_{\mathbb{1}} = x$

(ii) for every (finite or infinite) partition $\mathcal{F} = (\theta_i)_{i\in I}$, the vector $\bigotimes_{i\in I} x_{\theta_i}$ exists in the sense of App.A.3 and its image under $\Lambda_{\mathcal{F}}$ is equal to $x_{\vee\theta_i}$.

Moreover the mappings $x \longmapsto x_\theta$ are continuous.

We take an element θ and we set

$$\xi = (\theta, \theta') \quad , \quad \xi' = (\theta', \theta) \; ;$$

since x is factorizable there exist uniquely determined elements x_1 and x_1' in H_θ , x_2 and x_2' in $H_{\theta'}$ satisfying

$$(x_1 | \omega_\theta) = (x_1' | \omega_\theta) = (x_2 | \omega_{\theta'}) = (x_2' | \omega_{\theta'}) = 1$$

$$\Lambda_\xi (x_1 \otimes x_2) = \Lambda_{\xi'} (x_2' \otimes x_1') = x \; ;$$

by axiom (iv) of def.5.1 we have $x_1 = x_1'$, $x_2 = x_2'$; hence we can set $x_\theta = x_1$, $x_{\theta'} = x_2$ and we have

$$\Lambda_\xi (x_\theta \otimes x_{\theta'}) = \Lambda_{\xi'} (x_{\theta'} \otimes x_\theta) = x \; .$$

We now consider a partition $\theta = \bigvee_{i\in I} \theta_i$ and we set $\theta_o = \theta'$, $I' = I \cup \{o\}$; we drop the letters $\Lambda_{\mathcal{F}}$ for notational simplification ; so we write

$$H = \bigotimes_{i\in I'}^{h\,(\omega_{\theta_i})} H_{\theta_i} \; .$$

Since x is factorizable we know by Prop.A.1 (App.A) that there exist elements $x^i \in H_{\theta_i}$ such that $\otimes x^i$ exists and is equal to x ; since $(x | \omega) = 1$ we can assume $(x^i | \omega_{\theta_i}) = 1$ and then the x^i are uniquely determined. For each $i \in I'$ we can write

$$H_{\theta_i} \overset{h}{\otimes} \Big(\overset{h}{\underset{j\in I'-i}{\otimes}}{}^{(\omega_{\theta_j})} H_{\theta_j} \Big) = H$$

$$x^i \otimes \Big(\underset{j\in I'-i}{\otimes} x^j \Big) = x$$

and also

$$x_{\theta_i} \otimes x_{\theta_i'} = x \ ;$$

this implies $x^i = x_{\theta_i}$. Similarly we can write

$$H_{\theta_o} \overset{h}{\otimes} \Big(\overset{h}{\underset{i\in I}{\otimes}}{}^{(\omega_{\theta_i})} H_{\theta_i} \Big) = H$$

$$x^o \otimes \Big(\underset{i\in I}{\otimes} x^i \Big) = x$$

$$x_{\theta'} \otimes x_\theta = x$$

implies

$$x_\theta = \underset{i\in I}{\otimes} x^i = \underset{i\in I}{\otimes} x_{\theta_i}$$

which proves assertion (ii).

The continuity of the mapping $x \longmapsto x_\theta$ follows from the following inequality, where we have set $x'_\theta = x_\theta - \omega_\theta$, $y'_\theta = y_\theta - \omega_\theta$:

$$\begin{aligned} \| x - y \|^2 &= \| x_\theta \otimes x_{\theta'} - y_\theta \otimes y_{\theta'} \|^2 \\ &= \| (\omega_\theta + x'_\theta) \otimes (\omega_{\theta'} + x'_{\theta'}) - (\omega_\theta + y'_\theta) \otimes (\omega_{\theta'} + y'_{\theta'}) \|^2 \\ &= \| \omega_\theta \otimes (x'_{\theta'} - y'_{\theta'}) + (x'_\theta - y'_\theta) \otimes \omega_{\theta'} + x'_\theta \otimes x'_{\theta'} - y'_\theta \otimes y'_{\theta'} \|^2 \\ &\geq \| x_\theta - y_\theta \|^2 \end{aligned}$$

(note that $(x'_\theta | \omega_\theta) = (y'_\theta | \omega_\theta) = 0$).

<u>Lemma</u> 5.2. Let x , y be two elements of H^o and define x_θ , y_θ as in lemma 5.1 ; there exists a unique σ-additive complex function h on Θ satisfying $(x_\theta | y_\theta) = e^{h(\theta)}$ $\forall \theta$.

By lemma 5.1 and App.A.3 we have for every partition $\theta = \vee \theta_i$,

$$\sum |(x_{\theta_i} | y_{\theta_i}) - 1| < \infty$$

and

$$\prod (x_{\theta_i} | y_{\theta_i}) = (x_\theta | y_\theta) ;$$

in other words the function $\theta \longmapsto (x_\theta | y_\theta)$ is σ-multiplicative and our lemma follows from prop.C.1.

<u>Lemma</u> 5.3. We define x, y, x_θ , y_θ as in lemma 5.2 and we set

$$g(\theta) = |(x_\theta | y_\theta)|^2 / \|x_\theta\|^2 . \|y_\theta\|^2$$

$$f(\theta) = 1 - g(\theta) .$$

Then f is positive, non decreasing and for every partition $\theta = \vee \theta_i$ we have

$$f(\theta) \leq \sum f(\theta_i) < \infty . \qquad (5.1)$$

Moreover there exists for every $\varepsilon > 0$ a finite partition $\mathbb{1} = \vee \theta_i$ such that $f(\theta_i) \leq \varepsilon$ $\forall i$.

It is clear that $0 \leq g(\theta) \leq 1$; g is σ-multiplicative (same reasoning as in lemma 5.2) and therefore non increasing ; then f is positive and non decreasing ; moreover we can apply part b) of the proof of prop.C.1 with f replaced by g ; φ becomes our f and (5.1) is proved. The last assertion of our statement follows from lemma C.2.

<u>Proof of theorem</u> 5.1.

a) We choose a set K with a bijection $U : K \longrightarrow H^o$. We define the 0 element of K as $U^{-1}(\omega)$. By lemma 5.1 with every $a \in K$ we can associate a family (x_θ) satisfying (i) and (ii) with $x = Ua$; we set $x'_\theta = x_\theta - \omega_\theta$. We define the subsets K_θ by $a \in K_\theta$ iff $x'_{\theta'} = 0$; clearly (K_θ) is a non decreasing family and we have $0 \in K_\theta \ \forall \theta$, $K_0 = \{0\}$, $K_1 = K$.

b) We now define a function $(a,b) \longmapsto (a|b)$ by $(a|b) = h(1)$ with the notations of lemma 5.2 where $x = Ua$, $y = Ub$. We have the following properties :

- $(Ua|Ub) = e^{(a|b)}$
- $(a|b) = 0$ if $a = 0$ since $x = \omega$ implies $x_\theta = \omega_\theta$, $(x_\theta|y_\theta) = 1$, $h(\theta) = 0$.
- $(a|a) \geqslant 0$ for $(x_\theta|x_\theta) \geqslant 1$ (recall that $(x_\theta|\omega_\theta) = 1 = \|\omega_\theta\|^2$).
- $(a|a) = 0 \Longrightarrow a = 0$ since $h(1) = 0$ implies $(x|x) = 1$, $x = \omega$.
- $(a|b) = \overline{(b|a)}$ (immediate)
- $a \in K_\theta$, $b \in K_{\theta'} \Longrightarrow (a|b) = 0$ since for every $\eta \in \Theta$ we have $\eta = \eta_1 \vee \eta_2$ where $\eta_1 = \eta \wedge \theta$, $\eta_2 = \eta \wedge \theta'$ and

$$(x_\eta|y_\eta) = (x_{\eta_1}|y_{\eta_1})\,(x_{\eta_2}|y_{\eta_2})$$
$$= (x_{\eta_1}|\omega_{\eta_1})\,(\omega_{\eta_2}|y_{\eta_2}) = 1 .$$

- $(a|b) = 0 \ \forall b \in K_\theta \Longrightarrow a \in K_{\theta'}$. In fact for every $y \in H^o$ of the form $y_\theta \otimes \omega_{\theta'}$ we have

$$1 = e^{(a|b)} = (x|y) = (x_\theta|y_\theta)(x_{\theta'}|\omega_{\theta'}) = (x_\theta|y_\theta)$$

$$0 = (x_\theta - \omega_\theta|y_\theta) ;$$

then for every $z = z_\theta \otimes z_{\theta'} \in H^o$

$$((x_\theta - \omega_\theta) \otimes \omega_{\theta'}|z) = (x_\theta - \omega_\theta|z_\theta) = 0 ;$$

since H^o is total in H this implies $(x_\theta - \omega_\theta) \otimes \omega_{\theta'} = 0$, $x_\theta = \omega_\theta$, $a \in K_{\theta'}$.

We think that the present construction of the scalar product in K is simpler than Araki's ; but unfortunately we have not been able to simplify the construction of the linear structure given in c).

c) We now endow K with a linear structure, i.e. we define $\alpha a + \beta b$ for $\alpha, \beta \in \underline{C}$, $a, b \in K$. Set $x = Ua$, $y = Ub$ and define x_θ, x'_θ , y_θ , y'_θ as in part a).

c.1) We first prove that the element $z_{\mathcal{F}} = \bigotimes_{i \in I} (\omega_{\theta_i} + \alpha x'_{\theta_i} + \beta y'_{\theta_i})$ of H has a limit in H where $\mathcal{F} = (\theta_i)_{i \in I}$ runs over the set of all finite partitions of $\mathbb{1}$. We have to prove that for every $\varepsilon > 0$ there exists a finite partition $\mathcal{F}$ such that for every finite partition $\mathcal{G}$ finer than $\mathcal{F}$ we have

$$\| z_{\mathcal{F}} - z_{\mathcal{G}} \| \leq \varepsilon \quad . \tag{5.4}$$

For the sake of simplicity we take $\beta = 0$. Set

$$k = \max(\|x\|^2 , \|x\|^{4|\alpha|^2}) ; \tag{5.5}$$

choose ε' such that

$$k\left[\exp(\varepsilon' |\alpha|^2 \|x\|^4 \log\|x\|^2) - 1\right] \leq \varepsilon^2/4 \tag{5.6}$$

$$k\left[\exp(\varepsilon' |\alpha|^4 \|x\|^4 \log\|x\|^2) - 1\right] \leq \varepsilon^2/4 \tag{5.7}$$

and small enough in order that

$$t \leq \varepsilon' \|x\|^2 \Longrightarrow e^t - 1 - t \leq t^2 . \qquad (5.7')$$

Taking the notations of lemma 5.3 with $y = \omega$ we have

$$f(\theta) = 1 - \|x_\theta\|^{-2} = \|x'_\theta\|^2 / \|x_\theta\|^2 \quad ;$$

since $\|x\| \geq 1$, $\|x_\theta\| \geq 1$, $\|x_\theta\| \leq \|x\|$ we have

$$f(\theta) \geq \|x'_\theta\|^2 / \|x\|^2 . \qquad (5.8)$$

By lemma 5.3 there exists a finite partition of $\mathbb{1} : \mathcal{F} = (\theta_i)_{i\in I}$ such that $f(\theta_i) \leq \varepsilon' \ \forall i \in I$; we shall write ω_i , x_i , x'_i instead of ω_{θ_i} , x_{θ_i} , x'_{θ_i} . Setting $z_i = \omega_i + \alpha x'_i$ we have $z_{\mathcal{F}} = \otimes z_i$.

Let $\mathcal{G}$ be a finite partition of $\mathbb{1}$ finer than $\mathcal{F}$: $\mathcal{G} = (\theta_{i,j})$ where i runs over I , j runs over some set J_i , and $\theta_i = \vee_j \theta_{i,j}$; we have

$$z_{\mathcal{G}} = \bigotimes_{i,j} (\omega_{i,j} + \alpha x'_{ij}) = \bigotimes_{i,j} z_{ij}$$

$$\omega_i = \bigotimes_j \omega_{ij}$$

$$x_i = \bigotimes_j x_{ij}$$

$$x'_i = x_i - \omega_i = \bigotimes_j (\omega_{ij} + x'_{ij}) - \bigotimes_j \omega_{ij}$$

$$= u_i + v_i \qquad (5.9)$$

where u_i is the sum of those terms in $\bigotimes_j (\omega_{ij} + x'_{ij})$ which contain exactly one x'_{ij} and v_i the sum of those which contain several ones. Then

$$z_i = \omega_i + \alpha x_i' = \omega_i + \alpha u_i + \alpha v_i$$

$$\bigotimes_j z_{ij} = \bigotimes_j (\omega_{ij} + \alpha x_{ij}')$$

$$= \omega_i + \alpha u_i + w_i$$

where w_i is the sum of those terms in $\bigotimes_j (\omega_{ij} + \alpha x_{ij}')$ which contain several $\alpha x_{ij}'$;

$$z_{\mathcal{F}} = \bigotimes_i z_i = \bigotimes_i (\omega_i + \alpha u_i + \alpha v_i) \tag{5.10}$$

$$z_{\mathcal{G}} = \bigotimes_i \bigotimes_j z_{ij} = \bigotimes_i (\omega_i + \alpha u_i + w_i) .$$

Set $u = \bigotimes_i (\omega_i + \alpha u_i)$. We shall prove that

$$\| z_{\mathcal{F}} - u \| \leq \varepsilon/2 \tag{5.11}$$

$$\| z_{\mathcal{G}} - u \| \leq \varepsilon/2 \tag{5.12}$$

which will imply (5.4). We have by (5.8)

$$\| x_i' \|^2 \leq \varepsilon' \| x \|^2 \tag{5.13}$$

$$\begin{aligned} \Sigma \| x_i' \|^4 &\leq \varepsilon' \| x \|^2 . \Sigma \| x_i' \|^2 \\ &\leq \varepsilon' \| x \|^4 . \Sigma f(\theta_i) \\ &= - \varepsilon' \| x \|^4 . \Sigma (g(\theta_i) - 1) \\ &\leq - \varepsilon' \| x \|^4 . \Sigma \log g(\theta_i) \\ &= - \varepsilon' \| x \|^4 . \log \Pi g(\theta_i) \\ &= - \varepsilon' \| x \|^4 . \log g(\mathbb{1}) \\ &= \varepsilon' \| x \|^4 . \log \| x \|^2 \end{aligned} \tag{5.14}$$

Since $(\omega_{ij} | x_{ij}')$ is null, ω_i, u_i, v_i are mutually orthogonal and by (5.9) we have

$$\| u_i \|^2 \leq \| x_i' \|^2 \tag{5.15}$$

$$\| v_i \|^2 = \sum_{j_1 < j_2} \| x'_{i,j_1} \|^2 \cdot \| x'_{i,j_2} \|^2 + \sum_{j_1 < j_2 < j_3} \| x'_{i,j_1} \|^2 \cdot \| x'_{i,j_2} \|^2 \cdot$$

$$\| x'_{i,j_3} \|^2 + \ldots \prod_{j \in J_i} \| x'_{ij} \|^2$$

$$\leq (2!)^{-1} \Big(\sum_j \| x'_{ij} \|^2 \Big)^2 + \ldots ((\text{card } J_i)!)^{-1} \Big(\sum_j \| x'_{ij} \|^2 \Big)^{\text{card } J_i}$$

$$\leq \sum_{p=2}^{\infty} (p!)^{-1} \Big(\sum_j \| x'_{ij} \|^2 \Big)^p . \tag{5.16}$$

On the other hand

$$\| x_i' \|^2 = \| x_i \|^2 - 1 = \prod_j \| x_{ij} \|^2 - 1$$

$$= \prod_j (1 + \| x'_{ij} \|^2) - 1 \geq \sum_j \| x'_{ij} \|^2$$

and (5.16) implies

$$\| v_i \|^2 \leq \sum_{j=2}^{\infty} (p!)^{-1} \| x_i' \|^{2p}$$

$$= \exp(\| x_i' \|^2) - 1 - \| x_i' \|^2 \quad ;$$

by (5.13) and (5.7') this implies

$$\| v_i \|^2 \leq \| x_i' \|^4 . \tag{5.17}$$

We now turn to the proof of (5.11) ; we have

$$\| z_{\mathcal{F}} - u \|^2 = \| z_{\mathcal{F}} \|^2 + \| u \|^2 - 2 \operatorname{Re} (z_{\mathcal{F}} \mid u)$$

$$= \prod_i \| \omega_i + \alpha u_i + \alpha v_i \|^2 + \prod_i \| \omega_i + \alpha u_i \|^2$$

$$- 2 \operatorname{Re} \prod_i (\omega_i + \alpha u_i + \alpha v_i \mid \omega_i + \alpha u_i)$$

$$= \prod_i (1 + |\alpha|^2 \|u_i\|^2 + |\alpha|^2 \|v_i\|^2) - \prod_i (1 + |\alpha|^2 \|u_i\|^2)$$

$$= \prod_i (1 + |\alpha|^2 \|u_i\|^2) \left[\prod_i \frac{1 + |\alpha|^2 \|u_i\|^2 + |\alpha|^2 \|v_i\|^2}{1 + |\alpha|^2 \|u_i\|^2} - 1 \right] \quad (5.18)$$

by (5.15) we have

$$1 + |\alpha|^2 \|u_i\|^2 \leq 1 + |\alpha|^2 \|x_i'\|^2$$

$$\leq \begin{cases} 1 + \|x_i'\|^2 & \text{if } |\alpha| < 1 \\ (1 + \|x_i'\|^2)^2 |\alpha|^2 & \text{if } |\alpha| \geq 1 \end{cases}$$

$$= \begin{cases} \|x_i\|^2 & \text{if } |\alpha| < 1 \\ \|x_i\|^4 |\alpha|^2 & \text{if } |\alpha| \geq 1 \end{cases}$$

whence by (5.5)

$$\prod_i (1 + |\alpha|^2 \|u_i\|^2) \leq k .$$

Then (5.18) becomes

$$\|z_{\mathcal{F}} - u\|^2 \leq k \left[\exp \sum_i \log \frac{1 + |\alpha|^2 \|u_i\|^2 + |\alpha|^2 \|v_i\|^2}{1 + |\alpha|^2 \|u_i\|^2} - 1 \right]$$

$$\leq k \left[\exp \sum_i \left(\frac{1 + |\alpha|^2 \|u_i\|^2 + |\alpha|^2 \|v_i\|^2}{1 + |\alpha|^2 \|u_i\|^2} - 1 \right) - 1 \right]$$

$$= k \left[\exp \sum_i \frac{|\alpha|^2 \|v_i\|^2}{1 + |\alpha|^2 \|u_i\|^2} - 1 \right]$$

$$\leq k \left[\exp \sum_i |\alpha|^2 \|v_i\|^2 - 1 \right] ;$$

by (5.17), (5.14) and (5.6) we get

$$\| z_{\mathcal{F}} - u \|^2 \leq k \left[\exp \sum_i |\alpha|^2 \| x_i' \|^4 - 1 \right]$$

$$\leq k \left[\exp |\alpha|^2 \varepsilon' \| x \|^4 \log \| x \|^2 - 1 \right] \leq \varepsilon^2/4$$

which proves (5.11). The proof of (5.12) is quite similar : one has to replace αv_i by w_i, which leads to replace $|\alpha|^2 \| x_i' \|^4$ by $|\alpha|^4 \| x_i' \|^4$ and to use (5.7) instead of (5.6).

c.2) We now have to prove that the element $z = \lim_{\mathcal{F}} z_{\mathcal{F}}$ belongs to H^o ; since $(z | \omega) = \lim (z_{\mathcal{F}} | \omega) = 1$ we have only to prove that z is factorizable. Let $\mathcal{G}$ be a finite partition of $\mathbb{1}$; if $\mathcal{F}$ is finer than $\mathcal{G}$, $z_{\mathcal{F}}$ is decomposable with respect to $\mathcal{G}$ since it is decomposable with respect to $\mathcal{F}$; and z is also decomposable with respect to $\mathcal{G}$ by prop.A.2.

We thus can define $\alpha a + \beta b$ as $U^{-1}z$; one then proves that K is a prehilbert space by computations similar to the preceding ones. It is clear that the element $0 = U^{-1}\omega$ defined in part a) is actually the 0 element of the vector space K.

d) The mapping U is bicontinuous since we have

$$\| Ua - Ub \|^2 = \exp(\| a \|^2) + \exp(\| b \|^2) - 2 \operatorname{Re} \exp((a | b)) \tag{5.19}$$

$$\| a - b \|^2 = \log \| Ua \|^2 + \log \| Ub \|^2 - 2 \log |(Ua | Ub)| .$$

The space K is complete : in fact take a Cauchy sequence and set $x_n = Ua_n$; a_n has a limit a in the completion of K ; then by (5.19)

$$\| Ua_n - Ua_m \|^2 \longrightarrow \exp(\| a \|^2) + \exp(\| a \|^2) - 2 \operatorname{Re} \exp(\| a \|^2) = 0 ;$$

Ua_n has a limit x which lies in H^o since H^o is closed ; setting $x = Ua'$ we have $a' = \lim a_n$ by the continuity of U^{-1}.

e) Let us now prove that the K_θ's form a Boolean algebra of direct sum decompositions. First they are linear subspaces for

$$a \in K_\theta \Longleftrightarrow (a|b) = 0 \quad \forall\, b \in K_{\theta'}$$

(see part b)) ; they are closed because of the continuity of U and of the various mappings $x \longmapsto x_\theta$ (lemma 5.1). Moreover we already know that $K_{\theta'} = K_\theta^\perp$; it remains to prove that for each (finite or infinite) partition $\theta = \bigvee_{i\in I} \theta_i$, K_θ is the Hilbert sum of the K_{θ_i}'s. It is enough to prove that

$$a \in K_\theta \ , \ a \in K_{\theta_i}^\perp = K_{\theta'_i} \quad \forall\, i \Longrightarrow a = 0 \ ;$$

but

$$a \in K_\theta \Longrightarrow x_{\theta'} = \omega_{\theta'}$$

$$a \in K_{\theta'_i} \Longrightarrow x_{\theta_i} = \omega_{\theta_i} \quad ;$$

according to lemma 5.1 we can write

$$H = H_{\theta'} \overset{h}{\otimes} \Big(\overset{h}{\underset{i\in I}{\otimes}}{}^{(\omega_{\theta_i})} H_{\theta_i} \Big)$$

$$x = x_{\theta'} \otimes \Big(\underset{i\in I}{\otimes} x_{\theta_i} \Big)$$

$$= \omega_{\theta'} \otimes \Big(\underset{i\in I}{\otimes} \omega_{\theta_i} \Big) = \omega$$

whence $a = 0$.

f) Finally we must construct the isomorphisms $\Phi_\theta : H_\theta \longrightarrow S\,K_\theta$. For $a, b \in K$ we have

$$(Ua|Ub) = e^{(a\ b)} = (EXP\ a|EXP\ b) ;$$

since the elements Ua (resp. EXP a) are total in H (resp. S K) there exists an isomorphism Φ of H onto S K carrying every Ua into EXP a ; in the same manner we get isomorphisms Φ_θ of H_θ onto $S\,K_\theta$ and the required conditions are trivially satisfied.

g) There remains to prove property (iii) of the theorem, which is equivalent to saying that $(x|\omega) \neq 0$ for every $x \in H^1$. To this aim we note that for $x \in H^1$ we can still choose x_θ in H_θ such that $x_\theta \otimes x_{\theta'}$ is proportional to x (see lemma 5.1), but here x_θ is determined up to a non zero scalar multiple ; nevertheless the function g of lemma 5.3 is well defined and σ-multiplicative, and prop.C.1 shows that $g(\theta) \neq 0 \ \forall \theta$.

This completes the proof of theorem 5.1.

We complete theorem 5.1 with the description of the factorizable unitary operators.

<u>Definition</u> 5.3. With the notations of definition 5.2, a bounded linear operator T in H is called <u>factorizable</u> if for each finite partition of $\mathbb{1}$: $\mathcal{F} = (\theta_i)_{i \in I}$ there exist bounded linear operators T_i in H_{θ_i} such that $\Lambda_{\mathcal{F}}$ carries $\otimes\, T_i$ into T.

<u>Theorem</u> 5.2. We keep the assumptions of theorem 5.1 and write $H = S\,K$, $H_\theta = S\,K_\theta$. Then the factorizable unitary operators

are exactly the operators $U_{A,b,c}$ defined in § 2.2 where moreover A preserves all subspaces K_θ .

Take $U_{A,b,c}$ as indicated and a finite partition of $\mathbb{1}$: $\mathcal{F} = (\theta_i)_{i\in I}$; A and b have components A_i and b_i in each K_{θ_i} ; choose numbers c_i's such that $c = \prod c_i$; then by prop.2.4, $\Lambda_{\mathcal{F}}$ carries $\otimes U_{A_i,b_i,c_i}$ into $U_{A,b,c}$.

Conversely let U be a factorizable unitary operator in H ; U and U^{-1} preserve the set H^1 of all vectors λ EXP a , $\lambda \in \underline{C}^*$, $a \in K$; by lemma 2.1, U is of the form $U_{A,b,c}$ and we have to prove that A preserves the subspaces K_θ. We can write

$$K = K_\theta \oplus K_{\theta'}$$

$$H = H_\theta \otimes H_{\theta'}$$

$$U = U_\theta \otimes U_{\theta'}$$

Let

$$a = a_\theta \oplus 0 \in K_\theta$$

$$\text{EXP } a = \text{EXP } a_\theta \otimes \text{EXP } 0$$

$$U(\text{EXP } a) = U_\theta(\text{EXP } a_\theta) \otimes U_{\theta'}(\text{EXP } 0) ;$$

on the other hand we have

$$U(\text{EXP } a) = c \exp(-\|b\|^2/2 - (Aa \mid b)).\text{EXP } (A a + b)$$

$$= k \text{ EXP}((Aa)_\theta + b_\theta) \otimes \text{EXP } ((Aa)_{\theta'} + b_{\theta'}) ;$$

$\text{EXP}((Aa)_{\theta'} + b_{\theta'})$ must be proportional to $U_{\theta'}(\text{EXP } 0)$; taking $a = 0$ we see that EXP $b_{\theta'}$ is also proportional to $U_{\theta'}(\text{EXP } 0)$; this implies that

$$(Aa)_{\theta'} + b_{\theta'} = b_{\theta'}$$

$$(Aa)_{\theta'} = 0 \quad \text{and} \quad Aa \in K_\theta .$$

Chapter 6. FACTORIZABLE POSITIVE DEFINITE FUNCTIONS ON CURRENT GROUPS

In this chapter we shall consider special positive definite functions on special groups ; these groups are called "current groups" for physical reasons ; in mathematically more convenient terms such a group could be called "the group of the Borel sections of a Borel field of groups" or "of a Borel fiber bundle of groups", but here all fibers are identical to the same group (presumably the results of this chapter can be generalized to more general situations). Now roughly speaking if we have a partition of the base our group G will be decomposed into a direct product $\prod G_i$; a function on G is called factorizable if in such a case it is a direct product of functions on the G_i's. The aim of this chapter is to determine the general form of the factorizable positive definite functions on G ; the results will be interpreted in chapter 8 in terms of continuous products.

§ 6.1. Definitions

(For the definition and properties of Borel spaces see [9] or [32])

We denote by G a separable locally compact group with neutral element e ; by T a standard Borel space and by $\mathcal{B}$ the σ-algebra defining its Borel structure ; by F the set of all Borel mappings $T \longrightarrow G$ which take only a finite number of values ; F is a group for pointwise multiplication, which is sometimes called "weak current group" (see [35]) ; its neutral element will be denoted by ε ; F will be endowed with the discrete topology.

For every S in $\mathcal{B}$ we define $F^{(S)}$ similarly to F by replacing T by S ; every f in $F^{(S)}$ can be extended to an element $\tilde{f}$ in F taking the value e outside of S ; then $F^{(S)}$ can be considered as a subgroup of F. If we have a partition $S = \bigcup_{i \in I} S_i$ with S, $S_i \in \mathcal{B}$, we can consider $F^{(S)}$ as a subgroup of the product group $\prod_{i \in I} F^{(S_i)}$ which contains the restricted product $\prod'_{i \in I} F^{(S_i)}$ (for the definition see lemma 3.1). The restriction of an $f \in F$ to a subset S is denoted by $f^{(S)}$. The elements $\chi_{g,S}$ defined by

$$\chi_{g,S}(t) = \begin{cases} g & \text{if } t \in S \\ e & \text{if not} \end{cases}$$

generate the group F.

We now consider a positive definite function φ on F ; we denote by $\varphi^{(S)}$ its restriction to $F^{(S)}$; we say that S is φ-<u>negligible</u> if $\varphi^{(S)}$ is identical to 1. We assume the following conditions :

(i) $\varphi(\varepsilon) = 1$

(ii) φ is <u>factorizable</u>, i.e. for every countable partition $S = \cup S_i$ where S, $S_i \in \mathcal{B}$, and for every $f \in F$, we have

$$\sum_i |\varphi^{(S_i)}(f^{(S_i)}) - 1| < \infty$$

and

$$\prod_i \varphi^{(S_i)}(f^{(S_i)}) = \varphi^{(S)}(f^{(S)}) .$$

(iii) for every partition $T = \cup T_i$ where $T_i \in \mathcal{B}$, the set of the non φ-negligible T_i's is countable.

(iv) every point in T is φ-negligible.

(v) the Hilbert space H_φ associated with φ is separable.

(vi) for every S in $\mathcal{B}$ the function $G \ni g \longmapsto \varphi(\chi_{g,S})$ is continuous.

Our next aim is to determine the general form of such functions φ

§ 6.2. Results

By condition (ii) every countable union of φ-negligible sets is again φ-negligible ; moreover it is clear that every subset of a φ-negligible set is φ-negligible ; thus we may consider the Boolean algebra Θ, quotient of $\mathcal{B}$ by the φ-negligible sets ; Θ is σ-complete and, by condition (iii), σ-decomposable ; therefore it is complete (see App.C). Finally by condition (iv) it is non atomic (the proof is quite similar to a classical one in measure theory). For a given f, $\varphi(f^{(S)})$ depends only on the equivalence class of S in Θ ; the associated function on Θ is σ-multiplicative and prop.C.1 shows that $\varphi(f) \neq 0 \ \forall f$.

We now consider H_φ, U_φ, ξ_φ, Λ_φ as in § 3.1 and we set $H = H_\varphi$, $U = U_\varphi$, $\omega = \xi_\varphi$, $M = \Lambda_\varphi$; and similarly $H^{(S)} = H_{\varphi(S)}$, $U^{(S)} = U_{\varphi(S)}$, $\omega^{(S)} = \xi_{\varphi(S)}$, $M^{(S)} = \Lambda_{\varphi(S)}$. The following lemma says that, roughly speaking, $H^{(S)}$, $U^{(S)}$, $\omega^{(S)}$ depend only on the equivalence class of S in Θ.

<u>Lemma</u> 6.1. If S and S' differ by a φ-negligible set there exists

an isomorphism $H^{(S)} \longrightarrow H^{(S')}$ carrying $\omega^{(S)}$ into $\omega^{(S')}$, $M^{(S)}(\delta_{f^{(S)}})$ into $M^{(S')}(\delta_{f^{(S')}})$ and $U^{(S)}(f^{(S)})$ into $U^{(S')}(f^{(S')})$ for every f in F .

We can assume $S \supset S'$ and $S - S' = S''$ φ-negligible ; then for every f in F we have

$$\varphi^{(S)}(f^{(S)}) = \varphi^{(S')}(f^{(S')})$$

which implies for f_1 and f_2 in F :

$$(M^{(S)}(\delta_{f_1^{(S)}}) \mid M^{(S)}(\delta_{f_2^{(S)}})) = \varphi^{(S)}((f_2^{(S)})^{-1} f_1^{(S)})$$

$$= \varphi^{(S')}((f_2^{(S')})^{-1} f_1^{(S')})$$

$$= (M^{(S')}(\delta_{f_1^{(S')}}) \mid M^{(S')}(\delta_{f_2^{(S')}}))$$

and the lemma follows.

QED

In the following we shall write H_θ and ω_θ instead of $H^{(S)}$ and $\omega^{(S)}$ where θ is the equivalence class of S in Θ. We now prove that the hypotheses of definition 5.1 are fulfilled. Consider a (necessarily countable) partition $\mathcal{F} : \theta = \bigvee_{i \in I} \theta_i$, take $S_i \in \theta_i$ and set $S = \cup S_i \in \theta$; we have

$$\prod{}' F^{(S_i)} \subset F^{(S)} \subset \prod F^{(S_i)} \quad ;$$

by condition (ii) above and lemma 3.1 there exists an isomorphism

$$\Lambda_{\mathcal{F}} : \overset{h}{\otimes}{}^{(\omega_{\theta_i})} H_{\theta_i} \longrightarrow H_\theta$$

carrying $\otimes \omega_{\theta_i}$ into ω_θ ; conditions (i) to (iv) of definition 5.1 are easily verified.

The hypotheses of th.5.1 are fulfilled too : we already know that Θ is non atomic ; on the other hand the vectors $M(\delta_f)$, $f \in F$, are total in H and factorizable ; moreover

$$(M(\delta_f) | \omega) = \varphi(f^{-1}) \neq 0$$

which proves that H^o is total in H. Thus we can apply th.5.1 and write

$$H = S K \quad , \quad \omega = \text{EXP } 0$$

$$H_\theta = S K_\theta \quad , \quad \omega_\theta = \text{EXP } 0 \ ;$$

moreover for every f in F, U(f) is factorizable and by th.5.2 we can write

$$U(f) = U_{A(f),b(f),c(f)} \tag{6.1}$$

$$\varphi(f) = c(f).\exp(-\|b(f)\|^2/2) \tag{6.1'}$$

where

- A is a unitary representation of F in K which preserves each K_θ
- b is a 1-cocycle for A
- c is a mapping of F into $\underline{U}$ satisfying

$$c(ff')/c(f).c(f') = \exp\left[i \operatorname{Im}(b(f) | A(f).b(f'))\right]. \tag{6.2}$$

This already proves that U <u>and</u> φ <u>are of type</u> (S).

Since H is separable (condition (v) above), K is separable too and we can desintegrate it with a measure μ on T :

$$K = \int_T^{\oplus} K_t . d\mu(t)$$

in such a manner that K_θ becomes the set of all vector $x = (x_t)$ satisfying $x_t = 0$ a.e. outside of any subset S belonging to θ ; note that every φ-negligible set is μ-negligible. Since A(f) preserves each K_θ , it is decomposable with components $A(f)_t$ in K_t ; similarly b(f) has components $b(f)_t$. In the next lemma we "localize" the representation A and the cocycle b.

Lemma 6.2. There exist for every $t \in T$ a (continuous unitary) representation A_t of G in K_t and a (continuous) 1-cocycle b_t for A_t such that for every f in F we have $A(f)_t = A_t(f(t))$ and $b(f)_t = b_t(f(t))$ for almost all t .

By condition (vi) above, $\varphi(g)$ is continuous with respect to g (here g stands for the constant function g) ; by the Gelfand-Segal construction we know that U(g) is continuous too ; by th.2.1, A(g) and b(g) are continuous. Since G is locally compact and separable we can construct representations A_t of G in K_t such that for $g \in G$ we have $A(g)_t = A_t(g)$ a.e.; similarly one constructs 1-cocycles b_t such that $b(g)_t = b_t(g)$ a.e. (the proof is given in Appendix I).

Consider now an element f in F ; it takes values $g_1, \ldots g_n$ on some subsets $T_1, \ldots T_n$; we can write

$$K = \oplus K_{\theta_i} \quad \text{where } \theta_i \text{ is the class of } T_i$$

$$SK = \otimes SK_{\theta_i}$$

$$U(f) = \otimes U^{(T_i)}(f^{(T_i)}) = \otimes U^{(T_i)}(g_i) \ ;$$

in the same way as above we can construct objects A^i, b^i, c^i for $F^{(T_i)}$ such that

$$U^{(T_i)}(h) = U_{A^i(h),b^i(h),c^i(h)} \qquad \forall\, h \in F^{(T_i)} \ ;$$

then

$$U(f) = \otimes U_{A^i(g_i),b^i(g_i),c^i(g_i)} \ .$$

On the other hand we have

$$U(f) = U_{A(f),b(f),c(f)} \ ;$$

by prop.2.4 this implies

$$A(f) \mid K_{\theta_i} = A^i(g_i)$$

$$b(f) \mid K_{\theta_i} = b^i(g_i) \ ;$$

we see that $A(f) \mid K_{\theta_i}$ and $b(f) \mid K_{\theta_i}$ depend only on $f^{(T_i)}$, and therefore are equal to $A(g_i) \mid K_{\theta_i}$ and $b(g_i) \mid K_{\theta_i}$; this implies

$$A(f)_t = A(g_i)_t = A_t(g_i) = A_t(f(t))$$

$$b(f)_t = b(g_i)_t = b_t(g_i) = b_t(f(t))$$

almost everywhere on T_i, hence almost everywhere on T.

QED

We now prove something similar for $c(f)$.

<u>Lemma</u> 6.3. There exist a morphism $\alpha : F \longrightarrow \underline{R}$ and for every t in T a real function β_t on G with the following properties :

a) for every g in G the mapping $S \longmapsto \alpha(\chi_{g,S})$ is a real measure on $(T, \mathcal{B})$ singular with respect to μ

b) for almost all t

$$\beta_t(gg') - \beta_t(g) - \beta_t(g') = \mathrm{Im}\,(b_t(g) \mid A_t(g).b_t(g')) \tag{6.3}$$

for all $g, g' \in G$

c) $$c(f) = \exp\left[i\,\alpha(f) + i\int_T \beta_t(f(t)).d\mu(t)\right] \tag{6.4}$$

We take $f = \chi_{g,S}$ in formula (6.1′) ; g being fixed $\varphi(\chi_{g,S})$ depends only on the class θ of S and the mapping $\theta \longmapsto \varphi(\chi_{g,S})$ is a σ-multiplicative function on Θ ; the same is clearly true for

$$\theta \longmapsto \exp(-\|b(f)\|^2/2) = \exp(-\int \|b_t(f(t))\|^2/2.d\mu(t))$$
$$= \exp(-\int_S \|b_t(g)\|^2/2.d\mu(t)) \ ;$$

then $\theta \longmapsto c(\chi_{g,S})$ is also a σ-multiplicative function ; by prop.C.1 there exists for every g a real measure γ_g such that $c(\chi_{g,S}) = \exp(i\,\gamma_g(S))$. Following [35] th.2.2 we denote by γ'_g and γ''_g the absolutely continuous and the singular parts of γ_g with respect to μ : $\gamma_g(S) = \gamma'_g(S) + \gamma''_g(S)$. By (6.2) we get

$$\exp\left[i(\gamma_{gg'}(S) - \gamma_g(S) - \gamma_{g'}(S))\right] =$$
$$\exp\left[i\ \mathrm{Im}\int_S (b_t(g) \mid A_t(g)\ b_t(g')).d\mu(t)\right] ;$$

by the unicity in prop.C.1 this implies

$$\gamma_{gg'}(S) - \gamma_g(S) - \gamma_{g'}(S) = \mathrm{Im}\int_S (b_t(g) \mid A_t(g)\ b_t(g')).d\mu(t) \ ;$$

the left hand side of this equality is equal to

$$\gamma'_{gg'}(S) - \gamma'_g(S) - \gamma'_{g'}(S) + \gamma''_{gg'}(S) - \gamma''_g(S) - \gamma''_{g'}(S) \ ;$$

since the first half of this expression is absolutely continuous and the second is singular with respect to μ , this implies

$$\gamma'_{gg'}(S) - \gamma'_g(S) - \gamma'_{g'}(S) = \operatorname{Im} \int_S (b_t(g) | A_t(g)\, b_t(g')).d\mu(t) \tag{6.5}$$

$$\gamma''_{gg'}(S) - \gamma''_g(S) - \gamma''_{g'}(S) = 0 \ ;$$

we see that $g \longmapsto \gamma''_g(S)$ is a morphism of G into $\underline{R}$; we obtain the morphism α of the lemma by defining $\alpha(f) = \sum_{i=1}^{n} \gamma''_{g_i}(S_i)$ if f takes the value g_i on a set S_i ; and assertion a) is proved. As for γ', since γ'_g is absolutely continuous with respect to μ , by the Radon-Nikodym theorem we can write

$$\gamma'_g(S) = \int_S \beta_t(g).d\mu(t)$$

where β_t is some real function ; now (6.5) formally yields (6.3) and the rigorous proof is rather analogous to that of proposition I.1 in Appendix I . As to (6.4) we have, if $f = g_1, \dots g_n$ on $S_1, \dots S_n$:

$$\begin{aligned} c(f) &= \prod_j c(\chi_{g_j, S_j}) = \exp(i \sum_j \gamma g_j(S_j)) \\ &= \exp(i \sum_j \gamma''_{g_j}(S_j) + i \sum_j \gamma'_{g_j}(S_j)) \\ &= \exp\left[i\, \alpha(f) + i \int \beta_t(f(t)).d\mu(t)\right]. \end{aligned}$$

QED

Putting together lemmas 6.2 and 6.3 we get

Theorem 6.1. Keeping notations and assumptions of § 6.1, φ is a positive definite function of type (S) corresponding to a representation of type (S) in a space SK ; moreover there exist

- a desintegration $K = \int_T^{\oplus} K_t . d\mu(t)$ with some positive measure μ on T
- a morphism α of F into $\underline{R}$ such that for each g in G the mapping $S \longmapsto \alpha(\chi_{g,S})$ is a real measure singular with respect to μ.
- for each t in T a representation A_t of G in K_t , a 1-cocycle b_t for A_t and a real function β_t having coboundary $(g,g') \longmapsto \mathrm{Im}(b_t(g) \mid A_t(g).b_t(g'))$

such that for every f in F

$$\varphi(f) = \exp\left[i\,\alpha(f) + i\int_T \beta_t(f(t)).d\mu(t) - \frac{1}{2}\int_T \| b_t(f(t))\|^2 d\mu(t)\right].$$

Example 6.1. If G is compact, as we have seen in § 3.2 there exist $\omega_t \in K_t$ such that $b_t(g) = A_t(g)\,\omega - \omega$ and the function $\mathrm{Im}(b_t(g) \mid A_t(g).b_t(g'))$ is the coboundary of $\mathrm{Im}(A_t(g)\,\omega_t \mid \omega_t)$; then $\beta_t(g) - \mathrm{Im}(A_t(g)\,\omega_t \mid \omega_t)$ is a real character, hence null ; for the same reason α is null ; finally by th.3.1 we get

$$\varphi(f) = \exp\left[\int (\psi_t(f(t)) - \psi_t(e)).d\mu(t)\right] \tag{6.6}$$

where ψ_t is the p.d.f. $\psi_t(g) = (A_t(g)\,\omega_t \mid \omega_t)$.

Example 6.2. If G is commutative, as shown in the proof of th.3.2, case 3, we have two parts :

1) in the non trivial part of A_t , $\mathrm{Im}(b_t'(g) \mid A_t(g).b_t'(g'))$ is the coboundary of $\mathrm{Im}\int(\langle \chi, g\rangle - 1 - i\,J(g,\chi)).d\nu_t(\chi)$ where ν_t is some measure on $\widehat{G}$

2) in the trivial part, $\mathrm{Im}(b''_t(g) \mid b''_t(g'))$ must be the coboundary of $\beta_t(g) - \mathrm{Im}\int (\langle \chi, g\rangle - 1 - i\, J(g,\chi)).d\nu_t(\chi)$ which we call $\lambda_t(g)$; this implies that it is both symmetric and anti-symmetric, hence null, and λ_t is a real character.

We finally obtain

$$\varphi(f) = \exp\Big[i\,\alpha(f) + i\int \lambda_t(f(t)).d\mu(t) - \int Q_t(f(t)).d\mu(t) + \iint (\langle \chi, g\rangle - 1 - i\, J(g,\chi)).d\nu_t(\chi).d\mu(t)\Big]$$

where Q_t is some positive quadratic form on G ; we can also write

$$\varphi(f) = \exp\Big[i\,\sigma(f) - \int Q_t(f(t)).d\mu(t) - \quad \ldots\Big]$$

where σ is a morphism of F into $\underline{R}$; in fact we have

$$\sigma(\chi_{g,S}) = \alpha(\chi_{g,S}) + \int_S \lambda_t(g).d\mu(t)$$

and we see that the members in the righthand side are respectively the singular and the absolutely continuous parts of the measure $S \longmapsto \sigma(\chi_{g,S})$.

§ 6.3. Study of other current groups

In this paragraph we suppose that T is a locally compact σ-compact topological space with its natural Borel structure, and we denote by F^o the set of all f in F which are equal to e outside of some compact subset. Clearly F^o is a subgroup of F and possesses properties analogous to those of F indicated at the beginning of § 6.1 , except that $\chi_{g,S}$ is defined only for relatively compact S. We can consider p.d.f. φ on F^o with conditions identical to (i)....(vi) above except that (vi) is replaced by

(vi') for every relatively compact S in $\mathcal{B}$, the function $G \ni g \longmapsto \varphi(\chi_{g,S})$ is continuous.

Then one can prove a theorem 6.1' similar to th.6.1, the only difference being that the measures $S \longmapsto \alpha(\chi_{g,S})$ are defined only for relatively compact S .

Another possible and actually interesting current group is the set $\Gamma_c(T,G)$ of all continuous mappings $f : T \longrightarrow G$ which are equal to e outside some compact subset. For this group we can construct factorizable p.d.f. in the following way : take a Radon measure ν on T ; let M be a complex continuous function on $T \times G$ such that $M(t,e) = 0 \quad \forall t$; for every f in $\Gamma_c(T,G)$ the function $t \longmapsto M(t,f(t))$ is continuous with compact support so that we can consider the function

$$\varphi(f) = \exp\left[\int M(t,f(t)).d\nu(t)\right] ; \tag{6.7}$$

φ satisfies axioms (i),(ii),(iii),(v),(vi'), and also (iv) if ν is non atomic ; then the question arises : when φ is positive definite ? The next proposition gives a partial answer ; but before stating it we notice that φ can be extended to F^o ; in fact if f is of the form $\chi_{g,S}$ with S relatively compact, the function $t \longmapsto M(t,f(t))$ is null outside of S and its restriction to S is also the restriction of the continuous function $t \longmapsto M(t,g)$ defined on $\bar{S}$, which proves that our function is ν-integrable ; now the same is clearly true for an arbitrary f in F^o and we can define $\tilde{\varphi}$ on F^o by (6.7) with $f \in F^o$. If φ is positive definite

$\widetilde{\varphi}$ has the same property by continuity, and th.6.1' could give some information about M ; but one can prove directly the

<u>Proposition</u> 6.1. In order that φ be positive definite it is sufficient that for every relatively compact S in $\mathcal{B}$ and every point t in S , the function $G \ni g \longmapsto \exp[\nu(S).M(t,g)]$ be positive definite. This condition is also necessary if $M(t,g)$ is independent of t .

(See [12] and [18])

First assertion : for any $f_1,\dots f_n \in \Gamma_c(T,G)$, $\varphi(f_q^{-1}f_p)$ is a limit of expressions of the form

$$\exp\left[\sum_{i-1}^{m} \nu(S_i).M(t_i,f_q(t_i)^{-1} f_p(t_i))\right];$$

the corresponding matrices are positive definite by assumption, and so does by continuity the matrix with coefficients $\varphi(f_q^{-1}f_p)$. Second assertion : follows from the fact that φ positive definite implies $\widetilde{\varphi}$ positive definite.

QED

If ν is non atomic the condition in prop.6.1 becomes : $g \longmapsto \exp(k\,M(t,g))$ is positive definite for every t in T and every positive real k ; or equivalently by prop.4.3 : $g \longmapsto M(t,g)$ is conditionally positive definite and hermitian . Taking $T = G = \underline{R}$ and μ = Lebesgue measure we get the following result :

<u>Proposition</u> 6.2. Let M be a complex continuous function on $\underline{R}$ null at 0 ; the function on $\Gamma_c(\underline{R},\underline{R})$: $f \longmapsto \exp\int M(f(t)).dt$ is positive definite iff M has the form given in example 4.1.

(See [12] and App.E)

Chapter 7. GAUSSIAN MEASURES ON TOPOLOGICAL VECTOR SPACES

§ 7.1. Definitions and general properties

Definition 7.1. Let E be a real topological vector space which is a separable $(\mathcal{L}\mathcal{F})$-space (see Appendix G) ; a normalized finite positive Borel measure on E' is called Gaussian if its Fourier transform $x \longmapsto \varphi(x) = \mu(\hat{x})$ is of the form $e^{-\frac{1}{2}Q(x)}$ where Q is a non degenerate positive definite continuous quadratic form on E . We denote by Φ the symmetric bilinear functional associated with Q .

In what follows we use freely the notations of App.G. For every F in $\mathcal{F}(E)$ we can consider the measure $\mu_F = \Lambda_F(\mu)$ on F' ; its Fourier transform is $\varphi_F = e^{-\frac{1}{2}Q|F}$, which proves that μ_F is Gaussian ; more precisely let $e_1, \dots e_n$ be a basis of F , and M the matrix of $Q|F$ on this basis : $M = (m_{jk})$ with $m_{jk} = \Phi(e_j, e_k)$; set $N = M^{-1}$; then μ_F is given by

$$d\mu_F(u) = (2\pi)^{-n/2} (\det N)^{\frac{1}{2}} \exp(-\tfrac{1}{2}\Sigma n_{jk}u_ju_k).du_1 \dots du_n \tag{7.1}$$

where $u_1, \dots u_n$ are the components of u in the basis dual to $e_1, \dots e_n$. In particular for $F = \underline{R}\,x$, $x \in E$, we have

$$d\mu_F(u) = (2\pi)^{-\frac{1}{2}} Q(x)^{-\frac{1}{2}} \exp(-\tfrac{1}{2} Q(x)^{-1} u^2)\, du$$

this implies

$$\mu(\tilde{x}^2) = \int u^2 \, d\mu_F(u) = Q(x)$$

whence by the polarization identity

$$\mu(\tilde{x}\tilde{y}) = \Phi(x,y) \qquad \forall\ x,y \in E . \tag{7.2}$$

Denoting by H the completion of E with respect to the scalar product Φ, we see that the mapping $x \longmapsto \tilde{x}$ extends to an isometric linear mapping of H into $L^2(E',\mu;\underline{R})$; we still write $\tilde{x}$ for this extension. If now $e_1,\ldots e_n$ belong to H we see by continuity that the image of μ under the mapping

$$E' \longrightarrow \underline{R}^n$$

$$\chi \longmapsto (e_1(\chi),\ldots e_n(\chi))$$

is the measure

$$(2\pi)^{-n/2} (\det N)^{\frac{1}{2}} \exp(-\tfrac{1}{2} \Sigma\, n_{jk}\, u_j\, u_k)\, du_1 \ldots du_n \tag{7.3}$$

where $N = (n_{jk}) = M^{-1}$, $M = (\Phi(e_j,e_k))$.

Therefore the random variables x are Gaussian ; if x and y are orthogonal, $\tilde{x}$ and $\tilde{y}$ are orthogonal, hence independent. The mapping $x \longmapsto \tilde{x}$ is a Gaussian linear process with covariance Φ.

Now we can write $E \subset H \subset E'$ and we have

Proposition 7.1. If E is infinite dimensional, H is μ-negligible.

Let B_r be the ball in H with center 0 and radius r ; for every integer n there exists an n-dimensional subspace F in E and an orthonormal basis $e_1,\ldots e_n$ of F ; we have $\Lambda_F(B_r) \subset B_{F,r}$ where $B_{F,r}$ is the ball in F' with radius r ; therefore, using formula (7.1) :

$$\mu(B_r) \leq \mu_F(B_{F,r}) = (2\pi)^{-n/2} \int_{B_{F,r}} \exp(-\|u\|^2/2).du$$

$$= (2\pi)^{-n/2}\ S_n \int_0^r e^{-t^2/2}\ t^{n-1}\, dt$$

$$\leqslant (2\pi)^{-n/2} S_n r^n/n$$

where $S_n = 2\pi^{n/2}/\Gamma(n/2)$ is the area of the unit sphere in $\underline{R}^n$; this being true for every n we have $\mu(B_r) = 0$ and $\mu(H) = 0$ since H is the union of the B_r.

Remark 7.1. Suppose we are given a mapping $t \longmapsto \xi_t$ of a set T into a real Hilbert space $\mathcal{H}$ such that the elements ξ_t are total in $\mathcal{H}$; set $K(s,t) = (\xi_s | \xi_t)$ $\forall$ s,t $\in$ T ; to every vector η in $\mathcal{H}$ we can associate the real function u_η on T defined by

$$u_\eta(t) = (\xi_t | \eta) ;$$

we have in particular

$$u_{\xi_s}(t) = K(s,t) \quad \text{i.e.} \quad u_{\xi_s} = K(s,.) .$$

Clearly the mapping u is linear and injective ; we thus obtain an isomorphism of $\mathcal{H}$ onto a vector space $\mathcal{H}'$ of real functions on T ; we can define a scalar product on $\mathcal{H}'$ by $(u_{\eta_1} | u_{\eta_2}) = (\eta_1 | \eta_2)$; in particular

$$(K(s,.) \mid K(t,.)) = K(s,t) ;$$

K is called a reproducing kernel for the Hilbert space $\mathcal{H}'$.

We now apply this construction with T = E , $\mathcal{H}$ = closed subspace of $L^2(E',\mu;\underline{R})$ generated by all $\tilde{x}$, $\xi_x = \tilde{x}$ $\forall x \in E$; then $K(x,y) = \Phi(x,y)$; moreover for every η in $\mathcal{H}$ there exists y in H such that $\eta = \tilde{y}$ and we have, for every x in E

$$u_\eta(x) = (\tilde{x} | \eta) = (\tilde{x} | \tilde{y}) = \Phi(x,y) ;$$

we thus see that the realization of $\mathcal{H}$ as a Hilbert space with reproducing kernel is nothing but the realization of H as a set of continuous linear functionals on E.

Gaussian measures as product measures

If we take a linearly independant countable subset total in E and if we apply to it the Schmidt orthonormalization process, we obtain an orthonormal basis $e_1, e_2, \dots$ of H with $e_n \in E \ \forall n$; let F be the linear subspace of E algebraically generated by e_1, $e_2, \dots$; we put on F the finest locally convex topology (i.e. the one for which every linear mapping of F in a locally convex vector space is continuous) ; since the injections $F \longrightarrow E$ and $E \longrightarrow H$ are continuous we can write

$$\underline{R}^{(\underline{N})} \sim F \subset E \subset H \subset E' \subset F' \sim \underline{R}^{\underline{N}} .$$

Now μ , considered as a measure on F', has the following Fourier transform :

$$\underline{R}^{(\underline{N})} \ni x = (x_n) \longmapsto e^{-\frac{1}{2}Q(x)} = e^{-\frac{1}{2}\sum x_n^2} = \prod e^{-\frac{1}{2}x_n^2}$$

which proves that μ , as a measure on F', is the product of the reduced Gaussian measures on the various components $\underline{R}$. In short :

Proposition 7.2. Every Gaussian measure can be considered, by enlarging the space on which it is defined, as a countable product of measures equal to the reduced Gaussian measure on $\underline{R}$.

§ 7.2. Relation with symmetric Hilbert spaces

We keep the notations and assumptions of § 7.1 and we set $(x|y) = \Phi(x,y)$, $\|x\|^2 = Q(x)$.

Theorem 7.1. There exists a unique isomorphism T of SH onto $L^2(E', \mu ; \underline{R})$ such that $T(\mathrm{EXP}\ a) = \exp(\tilde{a} - \|a\|^2/2)$ for every a in H. If $a_1, \dots a_k$ are mutually orthogonal non zero elements of H, we have

$$T\,(P_{n_1+\dots n_k}(a_1^{\otimes n_1} \otimes \dots \otimes a_k^{\otimes n_k})) =$$

$$((n_1+\dots+n_k)!)^{-\frac{1}{2}} \prod_{j=1}^{k} \| a_j \|^{n_j} .\, h_{n_j}(\tilde{a}_j / \| a_j \|)$$

where P_n is the orthogonal projection of $H^{\otimes n}$ onto S^nH and $h_1, h_2, \dots$ are the Hermite polynomials (see App.E). In particular

$$T(a^{\otimes n}) = (n!)^{-\frac{1}{2}} \| a \|^n . h_n(\tilde{a} / \| a \|) .$$

The unicity of T is clear since the elements EXP a are total in SH. As in the proof of prop.7.2 we choose an orthonormal basis $e_1, e_2, \dots$ of H with $e_n \in E$ and we consider μ as the product of measures μ_j equal to the reduced Gaussian measure on $R_j = \underline{R}$; as was shown in example 2.1 there exists an isomorphism T_j of $S(\underline{R}\, e_j)$ onto $L^2(R_j, \mu_j; \underline{R})$ carrying each EXP a , $a \in \underline{R}\, e_j$, into the function $u \longmapsto \exp(au - a^2/2)$. On the other hand we have by prop.2.3 an isomorphism

$$S\,H \longrightarrow \overset{h}{\otimes}{}^{(\mathrm{EXP}\ 0)}\, S(\underline{R}\, e_j)$$

$$\mathrm{EXP}\ a \longmapsto \otimes\ \mathrm{EXP}\ a_j \qquad \forall\ a = (a_j) \in F \quad ;$$

an isomorphism

$$\otimes\, T_j \ :\ \overset{h}{\otimes}{}^{(\mathrm{EXP}\ 0)}\, S(\underline{R}\, e_j) \longrightarrow \overset{h}{\otimes}{}^{(1)}\, L^2(R_j, \mu_j; \underline{R})$$

$$\otimes\ \mathrm{EXP}\ a_j \longmapsto \otimes \exp(\tilde{a}_j - a_j{}^2/2)$$

and finally by App.D.3 an isomorphism

$$\overset{h}{\otimes}{}^{(1)}\ L^2(R_j, \mu_j; \underline{R}) \longrightarrow L^2(E', \mu\ ; \underline{R})$$

carrying each element $\otimes f_j$ (where $f_j \in L^2(R_j, \mu_j; \underline{R})$ and $f_j = 1$ for almost all j) into the function $\chi \longmapsto \prod f_j(\chi_j)$.

We thus get an isomorphism T of SH onto $L^2(E', \mu ; \underline{R})$ carrying every EXP a with $a \in F$ into the function

$$\chi \longmapsto \prod_j \exp(\tilde{a}_j(\chi_j) - a_j{}^2/2) = \exp(\tilde{a}(\chi) - \|a\|^2/2)$$

and by continuity this still holds for $a \in H$.

Proof of the last assertion : for any real numbers $u_1, \dots u_k$ we have

$$\mathrm{EXP}(u_1 a_1 + \dots + u_k a_k) = \sum_{n=0}^{\infty} (n!)^{-\frac{1}{2}} . (u_1 a_1 + \dots + u_k a_k)^{\otimes n}$$

$$= \sum_{n=0}^{\infty} (n!)^{-\frac{1}{2}} \sum_{n_1 + \dots n_k = n} \frac{n!}{n_1! \dots n_k!} u_1^{n_1} \dots u_k^{n_k} . P_n(a_1^{\otimes n_1} \otimes \dots a_k^{\otimes n_k})$$

$$= \sum_{n_1 \dots n_k} \frac{((n_1 + \dots n_k)!)^{\frac{1}{2}}}{n_1! \dots n_k!} u_1^{n_1} \dots u_k^{n_k} . P_{n_1 + \dots n_k}(a_1^{\otimes n_1} \otimes \dots a_k^{\otimes n_k})$$

on the other hand

$$T(\mathrm{EXP}(u_1 a_1 + \dots u_k a_k)) = \exp\left[(u_1 a_1 + \dots u_k a_k)^{\sim} - \|u_1 a_1 + \dots u_k a_k\|^2/2\right]$$

$$= \exp\left[u_1 \tilde{a}_1 + \dots u_k \tilde{a}_k - (u_1^2 \|a_1\|^2 + \dots u_k^2 \|a_k\|^2)/2\right]$$

$$= \prod_{j=1}^{k} \exp(u_j \tilde{a}_j - u_j^2 \|a_j\|^2/2)$$

$$= \prod_{j=1}^{k} \sum_{n=0}^{\infty} \|a_j\|^n u_j^n (n!)^{-1} . h_n(\tilde{a}_j/\|a_j\|)$$

$$= \sum_{n_1 \dots n_k} \prod_{j=1}^{k} \|a_j\|^{n_j} u_j^{n_j} (n_j!)^{-1} . h_{n_j}(\tilde{a}_j/\|a_j\|) ;$$

then our assertion follows by identifying the coefficients of $u_1^{n_1} \dots u_k^{n_k}$ in both expressions of $T(EXP(u_1a_1 + \dots u_ka_k))$.

<u>Corollary</u> 7.1. If $e_1, e_2, \dots$ is an orthonormal basis of H, the elements

$$\left[\frac{(n_1 + n_2 + \dots)!}{n_1! n_2! \dots}\right]^{\frac{1}{2}} T\left[P_{n_1+n_2+\dots}(e_1^{\otimes n_1} \otimes e_2^{\otimes n_2} \otimes \dots)\right]$$

$$= \prod_{j=1}^{\infty} (n_j!)^{-\frac{1}{2}} h_{n_j}(\tilde{e}_j)$$

constitute an orthonormal basis of $L^2(E', \mu; \underline{R})$ when $(n_1, n_2, \dots)$ runs over all sequences of positive integers which are zero almost everywhere.

In fact the elements

$$\left[\frac{(n_1 + n_2 + \dots)!}{n_1! n_2! \dots}\right]^{\frac{1}{2}} P_{n_1+n_2+\dots}(e_1^{\otimes n_1} \otimes e_2^{\otimes n_2} \otimes \dots)$$

form an orthonormal basis of SH (see § 2.1).

QED

By remark 2.0 we can assert that SH_c is canonically isomorphic to $L^2(E', \mu; \underline{C})$ which is the complexification of $L^2(E', \mu; \underline{R})$; the following theorem gives further information about this isomorphism.

<u>Theorem</u> 7.2. The isomorphism T of theorem 7.1 extends to an isomorphism $SH_c \longrightarrow L^2(E', \mu; \underline{C})$ carrying, for every a in H_c, a into $\tilde{a}$ and EXP a into $\exp(\tilde{a} - \mu(\tilde{a}^2)/2)$.

($\tilde{a}$ is defined as $\tilde{a}_1 + i\,\tilde{a}_2$ if $a = a_1 + i\,a_2$, a_1 and $a_2 \in H$).

The proof is quite similar to that of th.7.1 ; the required isomorphism is the tensor product of the various isomorphisms $S(\underline{C}\, e_j) \longrightarrow L^2(R_j, \mu_j; \underline{C})$ described in example 2.1.

Remark 7.2. The preceding results, as well as many subsequent ones, can be transposed to the general situation of linear stochastic processes by using App.G.2. Suppose for instance that we are given a Gaussian linear process X on E with a probability space $(\Omega, \mathcal{A}, P)$ with continuous characteristic functional, and denote by $\mathcal{B}$ the sub-σ-algebra of $\mathcal{A}$ defined by the functions X_x, $x \in E$. Then th.7.2 yields an isomorphism of SH_c onto the space $L^2(\Omega, \mathcal{B}, P|\mathcal{B})$.

§ 7.3. Examples of Gaussian measures

Example 7.1. We take $E = \mathcal{D}(\underline{R}_+ ; \underline{R})$, the space of all real functions on $\underline{R}_+ = [0, +\infty[$ which have compact supports and are infinitely differentiable on $]0, +\infty[$ and infinitely differentiable on the right at 0 ; we endow it with the usual, inductive limit topology of the space $\mathcal{D}$. We consider the quadratic form Q on E defined by

$$Q(x) = \int_0^\infty \left[\int_u^\infty x(t)\, dt\right]^2 du$$

and the associated symmetric bilinear form

$$\Phi(x,y) = \int_0^\infty \left[\int_u^\infty x(t)\, dt . \int_u^\infty y(s)\, ds\right] du .$$

Since E is nuclear, $e^{-Q/2}$ is the Fourier transform of a measure μ on E' ; μ is called the <u>Wiener measure</u>. The space $\mathcal{E}$ of all real continuous functions on $\underline{R}$ vanishing at 0 can be considered as a subspace of E' if we associate with every f in $\mathcal{E}$ the linear functional on E : $x \longmapsto \int_0^\infty f(t)\, x(t)\, dt$. It can be proved that μ is concentrated on $\mathcal{E}$ (see e.g. [4]) ; therefore we can associate with every s in $\underline{R}$ a function δ_s defined μ-almost everywhere on E', namely $\delta_s(f) = f(s)\ \forall f \in \mathcal{E}$; we thus get a stochastic process on $\underline{R}$: $s \longmapsto \delta_s$; moreover one can prove that δ_s belongs to $L^2(E',\mu)$ and that the covariance of our stochastic process is given by $\mu(\delta_s \delta_t) = \mathrm{Min}(s,t)$ [the reader can easily make a formal verification of this fact by approximating δ_s by a sequence of functions belonging to E]. This stochastic process is called <u>Wiener process</u> because it has been introduced by N.Wiener in his mathematical description of the Brownian motion.

Now for every x in E we set $Ax = x'$ (the derivative of x) ; A is a bijective mapping on E and we have

$$Q(Ax) = \int_0^\infty x(u)^2\, du = \|x\|_2^2 ;$$

therefore A extends to an isomorphism of $L^2(\underline{R}_+ ; \underline{R})$ onto H, the completion of E for Q. Applying th.7.1 we obtain an isomorphism of $S(L^2(\underline{R}_+ ; \underline{R}))$ onto $L^2(E',\mu ;\underline{R})$; its restriction to the subspace $S^n(L^2(\underline{R}_+ ; \underline{R}))$ is called "n-th stochastic Wiener integral" and the image of this subspace is called "n-th homogeneous Wiener chaos" ; the decomposition of an arbitrary element of $L^2(E',\mu ;\underline{R})$ on the basis described in cor.7.1 is referred to as the "Cameron-Martin decomposition" (cf. [24] and [30] 7.3).

<u>Example</u> 7.2. We take $E = \underline{R}^{(\underline{Z})}$; then $E' = \underline{R}^{\underline{Z}}$ and we consider a Gaussian measure μ on E' which is invariant under the shift transformation $A : (Au)_n = u_{n-1} \quad \forall\ u = (u_n) \in E'$; its Fourier transform $e^{-Q/2}$ must be invariant under the analogous transformation in E, hence Q must be of the form

$$Q(x) = \sum_{p,q} a_{p-q}\, x_p\, x_q$$

where a is a real positive definite function on $\underline{Z}$. Let ν be the positive measure on $\underline{T}$ such that $a_n = \int_{\underline{T}} e^{int}\, d\nu(t) \quad \forall\ n \in \underline{Z}$; with every $x = (x_n) \in E_c$ we associate the function f_x on $\underline{T}$: $f_x(t) = \sum_n x_n\, e^{int}$; then we have $Q(x) = \nu(|f_x|^2)$ and the mapping $x \longmapsto f_x$ extends to an isomorphism of H_c onto $L^2(\underline{T}, \nu)$. Applying th.7.2 we get an isomorphism

$$L \ :\ S(L^2(\underline{T}, \nu)) \longrightarrow L^2(E', \mu)$$

carrying every sufficiently regular function h on $\underline{T}$ into the function on E' : $u = (u_n) \longmapsto \sum x_n u_n$ where the x_n's are the Fourier coefficients of h. Let V denote the multiplication operator in $L^2(\underline{T}, \nu)$ by e^{it} ; then it is easily seen that L carries the operator $U_{V,0,1} = (I, V, V^{\otimes 2}, \ldots)$ into the shift operator U in $L^2(E', \mu)$: $(U\varphi)(u) = \varphi(A^{-1}u)$; this result can be used to study ergodicity, mixing,etc. properties of U (see [42]).

§ 7.4. The Wiener transform

Let μ be a Gaussian measure on E'.

a) We begin with the case where $E = \underline{R}$ and μ is the reduced Gaussian measure. We denote by T the isomorphisme $S\underline{C} \longrightarrow L^2(\underline{R}, \underline{C})$ described in example 2.1 and by W' the unitary operator in $S\underline{C}$:

$$W' = U_{i,0,1} = (1, i, i^2, \ldots) ;$$

we have $W'(\mathrm{EXP}\ a) = \mathrm{EXP}\ ia \quad \forall a \in \underline{C}$. We set $W = T\,W'\,T^{-1}$; this is a unitary operator in $L^2(\underline{R}, \mu)$ called <u>Wiener transform</u> in [40].

We also have $W = Z^{-1}\,F\,Z$ where Z is the isomorphism $L^2(\underline{R}, \ ) \longrightarrow L^2(\underline{R})$ defined by

$$(Zf)(x) = (2\pi)^{-1/4}.\ e^{-x^2/4}.f(x)$$

and F is the usual Fourier transform :

$$(Ff)(y) = \tfrac{1}{2}\,\pi^{-\frac{1}{2}} \int_{-\infty}^{+\infty} e^{ixy/2}.f(x).dx$$

[To prove this it is sufficient to verify that $W\,T\,\mathrm{EXP}\ a = Z^{-1}F\,Z\,T\,\mathrm{EXP}\ a \quad \forall a \in \underline{C}$ and this is a straightforward computation].

Moreover W carries every function $e^{ax-a^2/2}$, $a \in \underline{C}$, into $e^{iax-a^2/2}$ and every h_n into $i^n h_n$ (this is clear from the first definition of W) ; every function x^n into $(i2^{\frac{1}{2}})^n h_n(x2^{-\frac{1}{2}})$ and every function $h_n(x2^{-\frac{1}{2}})$ into $(i2^{-\frac{1}{2}})^n x^n$; this can be seen from the following formal computation (which could be made rigorous) :

$$\sum_{n=0}^{\infty} a^n\, 2^{-n/2}\, (n!)^{-1}\, W(x^n) = W(\sum a^n\, 2^{-n/2}\, (n!)^{-1}\, x^n)$$
$$= W(e^{ax/\sqrt{2}})$$
$$= e^{a^2/4} . W(e^{ax/\sqrt{2} - a^2/4})$$
$$= e^{a^2/4} . e^{iax/\sqrt{2} + a^2/4}$$
$$= e^{iax/\sqrt{2} + a^2/2}$$
$$= \sum i^n\, a^n\, (n!)^{-1}\, h_n(x/\sqrt{2})$$

whence $2^{-n/2}\, W(x^n) = i^n\, h_n(x/\sqrt{2})$. On the other hand

$$\sum a^n\, (n!)^{-1}\, W(h_n(x/\sqrt{2})) = W(\sum a^n\, (n!)^{-1}\, h_n(x/\sqrt{2}))$$
$$= W(e^{ax/\sqrt{2} - a^2/2})$$
$$= W(e^{-a^2/4} . e^{ax/\sqrt{2} - a^2/4})$$
$$= e^{-a^2/4} . e^{iax/\sqrt{2} + a^2/4}$$
$$= e^{iax/\sqrt{2}}$$
$$= \sum i^n\, 2^{-n/2}\, (n!)^{-1}\, a^n\, x^n$$

whence $W(h_n(x/\sqrt{2})) = i^n\, 2^{-n/2}\, x^n$.

b) We now pass to the general case. We call μ_j, T_j, W'_j, W_j what was called μ, T, W', W in part a); by th.7.2 we have an isomorphism $T : SH_c \longrightarrow L^2(E', \mu)$; we can identify SH_c with $\overset{h}{\otimes}{}^{(EXP\ 0)}\ \underline{SC}$ and $L^2(E', \mu)$ with $\overset{h}{\otimes}{}^{(1)}\ L^2(\underline{R}, \mu_j)$, so that we can consider the unitary operators $W' = \otimes W'_j$ in $S\underline{H}_c$ and $W = \otimes W_j$ in $L^2(E', \mu)$ and we have $W = T\, W' T^{-1}$; W is again called <u>Wiener transform</u>. Note that $W' = U_{i,o,1} =$

$(I, i, i^2, \ldots)$. It is clear from its definition that W carries, for every a in H_c, the function $e^{\tilde{a} - \mu(\tilde{a}^2)/2}$ into $e^{i\tilde{a} + \mu(\tilde{a}^2)/2}$, $h_n(\tilde{a}/\|a\|)$ into $i^n h_n(\tilde{a}/\|a\|)$ if $a \in H$, $a \neq 0$; and also the function $\prod_j u_j^{n_j}$ into $\prod_j (i\, 2^{\frac{1}{2}})^{n_j} h_{n_j}(u_j\, 2^{-\frac{1}{2}})$ and the function $\prod_j h_{n_j}(u_j\, 2^{-\frac{1}{2}})$ into $\prod_j (i\, 2^{-\frac{1}{2}})^{n_j} u_j^{n_j}$ where $n_j = 0$ a.e. (we have denoted by $u = (u_j)$ an arbitrary element of E').

§ 7.5. Equivalence of Gaussian measures

Theorem 7.3 (J.Feldman) Let μ and μ' be two Gaussian measures on E', Q and Q' the corresponding quadratic forms on E, Φ the bilinear form associated with Q , H the completion of E with respect to Φ. Then μ and μ' are either disjoint or equivalent ; they are equivalent if and only if there exists a symmetric Hilbert-Schmidt operator U in H such that $Q'(x) - Q(x) = \Phi(Ux, x)$ $\forall x \in E$; in that case the Radon-Nikodym derivative of μ' with respect to μ is equal almost everywhere to the function $\prod_{j=1}^{\infty} (\lambda_j + 1)^{-\frac{1}{2}} . \exp(\frac{1}{2} \lambda_j (\lambda_j + 1)^{-1} . \tilde{e}_j^{\,2})$ where $e_1, e_2, \ldots$ are the eigenvectors of U (orthonormal for Φ) and $\lambda_1, \lambda_2, \ldots$ the corresponding eigenvalues. (See [30], prop.8.6)

Remark 7.3. In the case where $e_n \in E$ $\forall n$, the proof can be considerably simplified ; in fact μ and μ' can be considered as product measures on a common space F' (see prop.7.2) and we can apply Kakutani's theorem (see App.H.3).

Proof of theorem 7.3.

a) We first prove that if there exists an operator U with the quoted properties, μ' is absolutely continuous with respect to μ with the indicated Radon-Nikodym derivative. We have $1+\lambda_j = Q'(e_j) > 0$ whence $\lambda_j > -1 \ \forall j$; moreover $\sum \lambda_j^2 < \infty$ implies $\lambda_j < 1$ for almost every j, say for $j \geqslant m$. Set

$$\varphi_j = (\lambda_j+1)^{-\frac{1}{2}}.\exp(\tfrac{1}{2}\lambda_j(\lambda_j+1)^{-1}.\tilde{e}_j^{\,2})$$

and denote by μ_j the image of μ under the mapping $\tilde{e}_j$; by (7.3) we have for every real $k < \frac{1}{2}$:

$$\begin{aligned}\int \exp(k\,\tilde{e}_j^{\,2}).d\mu &= \int \exp(k\,t^2).d\,\mu_j(t)\\ &= (2\pi)^{-\frac{1}{2}}\int \exp((k-\tfrac{1}{2})t^2).dt\\ &= (1-2k)^{-\frac{1}{2}}\end{aligned}$$

whence

$$\mu(\varphi_j) = (\lambda_j+1)^{-\frac{1}{2}}.(1-\lambda_j(\lambda_j+1)^{-1})^{-\frac{1}{2}} = 1.$$

For $j \geqslant m$, $-1 < \lambda_j < 1$ implies $\lambda_j(\lambda_j+1)^{-1} < \frac{1}{2}$ and

$$\begin{aligned}\mu(\varphi_j^2) &= (\lambda_j+1)^{-1}.(1-2\lambda_j(\lambda_j+1)^{-1})^{-\frac{1}{2}}\\ &= (1-\lambda_j^2)^{-\frac{1}{2}}.\end{aligned}$$

For every $n \geqslant m$ we set

$$\psi_n = \prod_{j=m}^{n} \varphi_j ;$$

then since the e_j's are independent we have $\psi_n \in L^2(E',\mu)$ and

$$\mu(\psi_n^2) = \prod_{j=m}^{n} \mu(\varphi_j^2) = \prod_{j=m}^{n} (1-\lambda_j^2)^{-\frac{1}{2}} ;$$

since $\sum \lambda_j^2$ is finite the product $\prod_{j=m}^{\infty} (1 - \lambda_j^2)^{-\frac{1}{2}}$ has a finite value $\alpha > 0$ and $\mu(\psi_n^2) \nearrow \alpha$. If $n < p$ we have

$$\mu(\psi_n \psi_p) = \mu(\psi_n^2 . \prod_{j=n+1}^{p} \varphi_j)$$

$$= \mu(\psi_n^2) . \prod_{j=n+1}^{p} \mu(\varphi_j) = \mu(\psi_n^2) ;$$

therefore

$$\mu((\psi_p - \psi_n)^2) = \mu(\psi_p^2) + \mu(\psi_n^2) - 2\mu(\psi_n \psi_p)$$

$$= \mu(\psi_p^2) - \mu(\psi_n^2)$$

tends to zero as n and p tend to infinity ; the ψ_n's form a Cauchy sequence in $L^2(E', \mu)$, hence converges to some limit ψ in L^2 and a fortiori in L^1, and we have

$$\mu(\psi) = \lim \mu(\psi_n) = 1 .$$

Let us set

$$\omega = \psi . \prod_{j=1}^{m-1} \varphi_j ;$$

because of the independence of the $\tilde{e}_j$'s, ω is μ-integrable and $\mu(\omega) - 1$. The functions $\omega_n = \prod_{j=1}^{n} \varphi_j$ converge in mean to ω since

$$\mu(|\omega - \omega_n|) = \mu(|\prod_{j=1}^{m} \varphi_j . (\psi - \psi_n)|)$$

$$= \mu(|\psi - \psi_n|) \longrightarrow 0 ;$$

but we also have convergence a.e.; in fact denote by $\mathcal{A}_n$ the σ-algebra on E' defined by $\tilde{e}_1, \dots \tilde{e}_n$; then for $n < p$ and for every $\mathcal{A}_n$-measurable function f we have

$$\mu(\omega_p f) = \mu(\psi_{n+1} \dots \psi_p . \omega_n . f) = \mu(\omega_n . f)$$

due to the independence of $\psi_{n+1},\dots \; _p$ and $\omega_n . f$; hence by App.H.1 we have

$$E^{\mathcal{A}_n}(\omega_p) = \omega_n$$

$$E^{\mathcal{A}_n}(\omega) = \lim_p E^{\mathcal{A}_n}(\omega_p) = \lim_p \omega_n = \omega_n$$

and our assertion follows from the martingale theorem (see App.H.2). Therefore we have μ -almost everywhere

$$\omega = \prod_{j=1}^{\infty} \varphi_j = \prod_{j=1}^{\infty} (\lambda_j+1)^{-\frac{1}{2}}.\exp(\tfrac{1}{2} \lambda_j(\lambda_j+1)^{-1}.\tilde{e}_j^{\,2}).$$

It remains to prove that $\mu' = \omega.\mu$. We set $\nu = \omega\mu$; it is enough to show that $\nu(e^{i\tilde{x}}) = e^{-\frac{1}{2}Q'(x)}$ for every x in E or, by continuity, for every x of the form $x = \Sigma x_j e_j$ with $x_j = 0$ a.e., say for $j > q$; then

$$\begin{aligned}
\nu(e^{i\tilde{x}}) &= \nu(\exp i \Sigma x_j \tilde{e}_j) = \mu(\omega . \exp i \Sigma x_j e_j) \\
&= \lim_n \mu(\prod_{j=1}^{n} \varphi_j.\exp i \Sigma x_j \tilde{e}_j) \\
&= \lim_n \prod_{j=1}^{n} (\lambda_j+1)^{-\frac{1}{2}}. \mu(\exp(\tfrac{1}{2} \lambda_j(\lambda_j+1)^{-1} \tilde{e}_j^{\,2} - i x_j\tilde{e}_j)) \\
&= \lim_n \prod_{j=1}^{n} (\lambda_j+1)^{-\frac{1}{2}} \int_{-\infty}^{+\infty} \exp(\tfrac{1}{2}\lambda_j(\lambda_j+1)^{-1}t^2-ix_jt)d\mu_j(t) \\
&= \lim_n \prod_{j=1}^{n} \exp(-x_j^2 (\lambda_j+1)/2) \\
&= \exp(-\tfrac{1}{2} \Sigma (\lambda_j+1) x_j^2) = \exp(- \tfrac{1}{2} Q'(x)) .
\end{aligned}$$

b) We now prove that if μ and μ' are not disjoint there exists an operator U with the properties indicated in the statement.

For each finite dimensional subspace F of E we call μ_F , μ'_F the projections of μ and μ' on F' ; since they are equivalent we can consider $\psi_F = d\mu'_F / d\mu_F$; let K_F be the element of S^2F such that $Q'(x) - Q(x) = \Phi(K_F , x \otimes x)$ $\forall x \in F$; K_F admits a decomposition

$$K_F = \sum_{j=1}^{n} \gamma_j . f_j \otimes f_j$$

where $f_1, \dots f_n$ constitute an orthonormal basis of F, the γ_j's are real and $\sum \gamma_j^2 = \|K_F\|^2$. Since μ and μ' are not disjoint the Hellinger integral $\int \sqrt{d\mu . d\mu'}$ has a value $\beta > 0$ and for every F we have

$$\mu_F(\psi_F^{\frac{1}{2}}) = \int \sqrt{d\mu_F . d\mu'_F} \geq \beta$$

(see App.H.3) ; by the proof of part a) we have

$$\psi_F = \prod_{j=1}^{n} (\gamma_j+1)^{-\frac{1}{2}} . \exp(\tfrac{1}{2} \gamma_j(\gamma_j+1)^{-1} . \tilde{f}_j^{\,2}) \; ;$$

since μ_F is the reduced Gaussian measure on F' we have

$$\mu_F(\psi_F^{\frac{1}{2}}) = \prod_{j=1}^{n} (\gamma_j+1)^{-\frac{1}{4}} (2\pi)^{-n/2} \int_{-\infty}^{+\infty} \exp(\tfrac{1}{4}\gamma_j(\gamma_j+1)^{-1} t^2 - \tfrac{1}{2} t^2) . dt$$

$$= \left[\prod_{j=1}^{n} 4 \, (\gamma_j+1)(\gamma_j+2)^{-2} \right]^{\frac{1}{4}} \; ;$$

therefore

$$\beta^4 \leq \prod_{j=1}^{n} 4(\gamma_j+1)(\gamma_j+2)^{-2} = \prod_{j=1}^{n} (1 - \gamma_j^2(\gamma_j+2)^{-2}) \qquad (7.4)$$

(note that $\gamma_j > -1$ implies $\gamma_j^2(\gamma_j+2)^{-2} < 1$)

$$-\log \beta^4 \geq -\sum_{j=1}^{n} \log(1 - \gamma_j^2(\gamma_j+2)^{-2}) \geq \sum_{j=1}^{n} \gamma_j^2(\gamma_j+2)^{-2} \qquad (7.5)$$

Taking $F = \underline{R} f_j$ we see by (7.4) that

$$\beta^4 \leq 4(\gamma_j+1)(\gamma_j+2)^{-2}$$

which implies that γ_j is bounded by some constant A ; then (7.5) implies

$$\| K_F \|^2 = \sum_{j=1}^{m} \gamma_j^2 \leq - (A + 2)^2 \log \beta^4 .$$

If a finite dimensional subspace F_1 is included in another one F_2, K_{F_1} is the orthogonal projection of K_{F_2} onto S^2F_1 ; therefore K_F has a limit K in S^2H. Finally take x in E and $F_2 \supset F_1 = \underline{R}\, x$; we have

$$\begin{aligned} \Phi(K , x\otimes x) &= \lim_{F_2} \Phi(K_{F_2} , x\otimes x) \\ &= \lim_{F_2} \Phi(K_{F_1} , x\otimes x) = \Phi(K_{F_1} , x\otimes x) \\ &= Q'(x) - Q(x) \end{aligned}$$

and our assertion is proved by taking for U the Hilbert-Schmidt operator associated with K.

c) If μ and μ' are not disjoint, by part b) there exists an U with the properties mentioned in the statement, and by part a), μ' is absolutely continuous with respect to μ ; but by symmetry μ is also absolutely continuous with respect to μ', so that the theorem is completely proved.

<u>Corollary</u> 7.2. Let μ be a Gaussian measure on E', $x \longmapsto e^{-\|x\|^2/2}$ its Fourier transform, H the completion of E with respect to $\| \;\; \|$, A an injective continuous linear operator in E, μ' the image of μ

under ${}^{t}A$ (the linear operator dual to A). Then μ, μ' are equivalent or disjoint ; they are equivalent iff A extends to a continuous linear operator $\bar{A}$ in H such that $\bar{A}^{*}\bar{A} - I$ is Hilbert-Schmidt.

Proof : μ' is Gaussian since

$$\mathcal{F}\mu'(x) = \mathcal{F}\mu(Ax) = e^{-\|Ax\|^2/2} \;;$$

by th.7.3, μ and μ' are equivalent or disjoint ; if they are equivalent there exists a Hilbert-Schmidt operator U in H such that

$$Q'(x) = \|Ax\|^2 = \|x\|^2 + (Ux|x) \;;$$

therefore A is continuous for the topology of H and its extension $\bar{A}$ to H satisfies

$$(\bar{A}^{*}\bar{A}\,x|x) = \|\bar{A}\,x\|^2 = (x|x) + (U\,x|x)$$

i.e.

$$\bar{A}^{*}\bar{A} - I = U .$$

The converse is immediate.

Remark 7.4. In particular if E is infinite dimensional and A is a scalar c , μ and μ' are equivalent iff $c = \pm 1$; in other words Gaussian measures are not quasi-invariant under homotheties with ratio other than ± 1 .

Corollary 7.3. Let Q and Q' be two continuous non degenerate positive quadratic forms on E, V and V' the representations of E associated with the p.d.f. $e^{-Q/2}$ and $e^{-Q'/2}$. Then V and V' are equivalent iff Q and Q' satisfy the conditions of th.7.3.

As shown in § 7.1 there exists a dense subspace F in E which is nuclear for a stronger topology ; $e^{-Q/2} \mid F$ and $e^{-Q'/2} \mid F$ are Fourier transforms of measures μ and μ' on F' ; $V \mid F$ and $V' \mid F$ can be realized in $L^2(F',\mu)$ and $L^2(F',\mu')$ as multiplication operators by $e^{i\tilde{x}}$; now we have

$$V \sim V' \Longleftrightarrow V \mid F \sim V' \mid F \Longleftrightarrow \mu \sim \mu' .$$

§ 7.6. Quasi-invariance of Gaussian measures with respect to translations

We consider a Gaussian measure μ on E' and define Q, Φ, H as in § 7.1 and we write $E \subset H \subset E'$; for every u in E' we denote by T_u the translation in E' defined by $T_u(\chi) = \chi + u$.

Theorem 7.4. For u and v in E' the measures $T_u(\mu)$ and $T_v(\mu)$ are either equivalent or disjoint ; they are equivalent if and only if $u - v$ belongs to H.

Define $e_1, e_2, \ldots$ and F as in prop.7.2 and write $\mu = \otimes \mu_n$ where μ_n is the reduced Gaussian measure on $\underline{R}$; set $\mu' = T_u(\mu)$, $\mu'' = T_v(\mu)$; their Fourier transforms φ and φ' are given by

$$\varphi'(x) = \int e^{i<\chi+u,x>} . d\mu(\chi) = e^{i<u,x> - Q(x)/2}$$

$$= \prod e^{iu_n x_n - x_n^2/2}$$

$$\varphi''(x) = \prod e^{iv_n x_n - x_n^2/2}$$

for every $x = \sum x_n e_n \in F$; this proves that μ' and μ'' are respectively equal to $\otimes \mu_n'$ and $\otimes \mu_n''$ where

$$d\mu_n'(t) = (2\pi)^{-\frac{1}{2}} e^{-(t-u_n)^2/2} dt$$

$$d\mu_n''(t) = (2\pi)^{-\frac{1}{2}} e^{-(t-v_n)^2/2} dt \quad ;$$

then $\psi_n = d\mu_n'' / d\mu_n'$ is given by

$$\psi_n(t) = \exp(t(v_n - u_n) - (u_n^2 - v_n^2)/2)$$

and we have

$$\int \sqrt{d\mu_n' . d\mu_n''} = \int \psi_n^{\frac{1}{2}} . d\mu_n' = \exp(-(u_n - v_n)^2/8) .$$

The theorem now follows from Kakutani's theorem (See App.H.3).

Corollary 7.4. The measure μ is quasi-invariant under T_u iff u belongs to H ; in that case the Radon-Nikodym derivative of $T_u(\mu)$ with respect to μ is the function $e^{\tilde{u} - \|u\|^2/2}$.

We see that the infinite dimensional case is throughly different from the finite dimensional one : in the second case there is only one class of measures on E' which is quasi-invariant under E', namely the class of the Lebesgue measure ; in the first case there are uncountably many classes of measures which are quasi-invariant under H ; moreover it can be proved that none of them is quasi-invariant under E' (see [12], ch.IV, § 5).

Remark 7.4. Theorem 7.4 could suggest the conjecture that μ is carried by some residue class modulo H ; however this conjecture is false if H is infinite dimensional ; in fact every residue class L modulo H is negligible : this follows from prop.7.1 if L = H and from cor.7.5 if $L \neq H$.

Application to certain representations of H and H_c

It follows from cor.7.4 that for every x in H we can define a unitary operator $V_1(x)$ in $L^2(E',\mu)$ by

$$(V_1(x).f)(\chi) = \exp(-\|x\|^2 - \tilde{x}(\chi)).f(\chi - 2x) ;$$

on the other hand we have another unitary operator

$$(V_2(x).f)(\chi) = \exp(i\,\tilde{x}(\chi)).f(\chi) ;$$

for $x \in H_c$, $x = x_1 + ix_2$, x_1 and $x_2 \in H$ we set

$$(V(x).f)(\chi) = \exp(-\|x_1\|^2 - i(x_1|x_2) + \tilde{x}(\chi)).f(\chi - 2x_1)$$

(we recall that $\tilde{x} = \tilde{x}_1 + i\tilde{x}_2$) ; then a simple calculation shows that the isomorphism $T^{-1} : L^2(E',\mu) \longrightarrow SH_c$ described in th. 7.2 carries $V(x)$ into the operator $U_{I,x,1}$ defined in §2.2 ; recall that

$$U_{I,x,1}\,(\mathrm{EXP}\ a) = \exp(-\|x\|^2/2 - (a|x)).\mathrm{EXP}\ (a + x) .$$

We thus see that V is a projective continuous representation of H_c in $L^2(E',\mu)$ with multiplier $(x,x') \longmapsto \exp(i\ \mathrm{Im}(x|x'))$, and it is irreducible by prop. 2.5 ; by continuity its restriction to any dense linear subspace of H_c is irreducible too ; this yields the following :

Theorem 7.5. The measure μ is ergodic with respect to an arbitrary dense linear subspace of H acting by translations.

Let K be such a subspace, f a bounded measurable function such that for every u in K we have $f(\chi + u) = f(\chi)$ a.e.; the multiplication operator by f commutes with all $V(x)$, $x \in K$, hence is a scalar k and f is equal a.e. to k .

Corollary 7.5. An arbitrary measurable affine subspace of E' has measure 0 or 1, and necessarily ß if it does not contain H.

Let L be an affine subspace, L_o the linear subspace parallel to L . First suppose that L_o contain H ; then L is invariant under all translations by elements of H, therefore by th.7.5 it has measure 0 or 1. Now suppose that L_o does not contain H ; there exists a vector $u \in H$, $u \notin L_o$; the affine subspaces $L + \alpha u$ where $\alpha \in \underline{R}$ are mutually disjoint ; if $\mu(L)$ is strictly positive, so does $\mu(L + \alpha u)$ since μ is quasi-invariant under $T_{\alpha u}$, which is absurd since μ is finite. It remains to prove that L has measure 0 if it does not contain 0 ; in that case we have $L \cap (-L) = \emptyset$; moreover $\mu(-L) = \mu(L)$ since μ is symmetric about 0 ; hence $\mu(L) = \frac{1}{2}\mu(L \cup (-L)) \leq \frac{1}{2}$ which implies $\mu(L) = 0$.

Remark 7.5. The Wiener transform in $L^2(E',\mu)$ exchanges the representations V_1 and V_2 ; more precisely we have

$$W.V_1(x).W^{-1} = V_2(-x) \qquad \forall\ x \in H$$

as is easily seen in the space SH_c .

§ 7.7. Ergodicity of Gaussian measures with respect to rotations

Let E be a real separable $(\mathcal{LF})$-space, Q a non degenerate continuous positive quadratic form on E, G the group of all continuous invertible linear transformations in E preserving Q, A the set of all normalized c.p.d.f. on E invariant under G ; we assume that E is infinite dimensional.

For every real number $c \geq 0$, the function e^{-cQ} belongs to A ; for every normalized positive measure ν on $\underline{R}_+ = [0,+\infty[$, the function

$$x \longmapsto \varphi_\nu(x) = \int e^{-cQ(x)} d\nu(c)$$

is again an element of A [to prove the continuity it is enough to prove it on each E_n (see App.G) and this follows from Lebesgue's convergence theorem]. The mapping $\nu \longmapsto \varphi_\nu$ is injective ; in fact ν has a Laplace transform

$$\mathcal{L}\nu\,(u+iv) = \int e^{-c(u+iv)} d\nu(c)$$

which is holomorphic in the half plane $u > 0$, and we have $\varphi_\nu(x) = \mathcal{L}_\nu(Q(x))$; then $\varphi_{\nu_1} = \varphi_{\nu_2}$ implies $\mathcal{L}\nu_1 = \mathcal{L}\nu_2$, which implies $\nu_1 = \nu_2$. To prove that $\nu \longmapsto \varphi_\nu$ is surjective we need a few lemmas.

<u>Lemma</u> 7.1. If $x,y \in E$ and $Q(x) = Q(y)$, there exists a transformation in G carrying x in y .

We can suppose $Q(x) = Q(y) = 1$ and x non proportional to y ; as shown in § 7.1 there exists an orthonormal basis $e_1, e_2, \ldots$ in H such that $e_1 = x$, e_2 is a linear combination of x and y , and $e_n \in E \;\; \forall n$; we define an orthogonal operator T in H by

$$Tx = y\,,\quad Ty = x\,,\quad Te_n = e_n \quad \forall\, n > 2\,;$$

it remains to prove that $T(E) \subset E$. Every z in E can be written $z = \sum_{n=1}^{\infty} z_n e_n$ with $\sum z_n^2 < \infty$; since $z_1e_1 + z_2e_2$ belongs to E, the vector $\sum_{n=3}^{\infty} z_n e_n$ also belongs to E ; Tz is the sum of this vector and of a linear combination of e_1, e_2 ; hence Tz belongs to E.

__Lemma__ 7.2. For every complex function ψ on $\underline{R}$ such that the function $x \longmapsto \psi(Q(x))$ is positive definite we have the following properties :

(i) $\psi(t) \geq 0 \ \forall \ t \geq 0$

(ii) the function $x \longmapsto -\Delta_c^{(1)} \psi(Q(x))$ is positive definite for every $c > 0$.

(See [43], prop.21.3 ; we recall that

$$\Delta_c^{(n)} \psi(t) = (-1)^n \sum_{k=0}^{n} (-1)^k \binom{n}{k} \psi(kc+t) \ )$$

__Proof of__ (i). For every n there exist $x_1, \dots x_n$ in E such that $\Phi(x_i, x_j) = t\,\delta_{ij}$ where Φ is the symmetric bilinear form associated with Q ; then

$$0 \leq \sum_{ij} \psi(Q(x_i - x_j)) = n\,\psi(0) + n(n-1)\,\psi(2t)$$

$$\psi(2t) \geq -\psi(0)/(n-1)$$

and letting n tend to ∞ we get (i).

__Proof of__ (ii). We set $\omega(t) = \Delta_c^{(1)} \psi(t)$; we take $a_1, \dots a_n$ in $\underline{C}$ and $x_1, \dots x_n$ in E ; there exists x_o such that $\Phi(x_o, x_i) = c\,\delta_{oi}$ for $i = o, \dots n$. We set

$$y_k = \begin{cases} x_k \\ x_{k-n} + x_o \end{cases} \qquad b_k = \begin{cases} a_k & \text{if } k = 1, \dots n \\ -a_{k-n} & \text{if } k = n+1, \dots 2n . \end{cases}$$

We have

$$0 \leq \sum_{ij=1}^{2n} b_i \bar{b}_j \psi(Q(y_i - y_j))$$

$$= \sum_{ij=1}^{n} a_i \bar{a}_j \psi(Q(x_i - x_j)) - \sum_{ij=1}^{n} a_i \bar{a}_j \psi(Q(x_i - x_j - x_o)) -$$

$$- \sum_{ij=1}^{m} a_i \bar{a}_j \psi(Q(x_i + x_o - x_j)) + \sum_{ij=1}^{m} a_i \bar{a}_j \psi(Q(x_i - x_j))$$

$$= 2 \left[\sum_{ij=1}^{m} a_i \bar{a}_j \psi(Q(x_i - x_j)) - \sum_{ij=1}^{m} a_i \bar{a}_j \psi(Q(x_i - x_j) + c) \right]$$

$$= -2 \sum_{ij=1}^{m} a_i \bar{a}_j \psi(Q(x_i - x_j))$$

which proves (ii).

QED

We now prove the surjectivity of the mapping $\nu \longmapsto \varphi_\nu$. Let φ be an element of A ; by lemma 7.1, $\varphi(x)$ is a function of Q(x), say $\psi(Q(x))$; by lemma 7.2 (ii), the function $x \longmapsto (-1)^n \Delta_c^{(n)} \psi(Q(x))$ is positive definite for every n, and by (i) we have $(-1)^n \Delta_c^{(n)} \psi(Q(x)) \geqslant 0$; in other words ψ is completely monotonic ; by Bernstein's lemma ψ is of the form

$$\psi(t) = \int e^{-ct}.d\nu(c)$$

where ν is a normalized positive measure on $\underline{R}_+$ and the proof is complete.

Thus $\nu \longmapsto \varphi_\nu$ is a bijection of the set of all normalized positive measures on $\underline{R}$ onto A ; it preserves the barycentres and therefore the extreme points ; since the extreme normalized positive measures are exactly the Dirac measures we have the following result :

Theorem 7.6. The extreme points of A are exactly the functions e^{-cQ} where $c \in \underline{R}_+$; moreover every element of A can be written in a unique manner as $\int e^{-cQ}.d\nu(c)$ where ν is a normalized positive measure on $\underline{R}_+$.

(See [39] and [43], §21)

We now let G act on E' by transposition and we have :

<u>Corollary</u> 7.6. Every measure μ on E' such that $\mathcal{F}\mu = e^{-cQ}$ with $c \geqslant 0$ is invariant and ergodic under G . If E is nuclear the ergodic invariant normalized measures on E' are exactly the measures $\mu_o = \delta_o$ and the Gaussian measures μ_c defined by $\mathcal{F}\mu_c = e^{-cQ}$ where $c > 0$; moreover every invariant normalized measure can be written in a unique manner as $\int \mu_c . d\nu(c)$ where ν is a normalized positive measure on $\underline{R}_+$.

In fact the ergodic invariant normalized measures are the extreme points in the set of all invariant normalized measures ; and if E is nuclear every c.p.d.f. is the Fourier transform of a measure.

<u>Remark</u> 7.6. In particular every Gaussian measure μ on E' is invariant and ergodic under the group of all continuous linear transformations in E preserving the covariance of μ ; in the case where $E = \underline{R}^{(\underline{N})}$ and $Q(x) = \sum x_n^2$, this fact also follows from [21]: since μ is of the form $\otimes \mu_n$ where the μ_n's are identical, μ is ergodic for a group which is much smaller than G , namely the group of all permutations of $\underline{N}$.

Chapter 8. CONTINUOUS PRODUCTS

§ 8.1. Introduction

The present chapter is an attempt to make rigorous the following formal considerations.

a) Continuous products of complex numbers

Given a complex function f on a set T with a measure μ , the continuous product of the function f (or of the family of the complex numbers f(t), $t \in T$) is defined as $\exp(\int \log f(t).d\mu(t))$; it is denoted by $\prod f(t)^{d\mu(t)}$ or $\prod f^{d\mu}$ or $\prod f$ and has the following properties :

$$\prod (f_1 f_2) = \prod f_1 . \prod f_2 \tag{8.1}$$

$$\prod \bar{f} = \overline{\prod f} \tag{8.2}$$

$$\prod |f| = |\prod f| \tag{8.3}$$

b) Continuous tensor products of Hilbert spaces

Given a family of Hilbert spaces H_t , $t \in T$, their continuous tensor product $\otimes H_t$ is generated by particular elements $\otimes x_t$, called decomposable, where the family (x_t) runs over some subset of the cartesian product $\prod H_t$; these elements have the following properties :

$$(\otimes x_t | \otimes y_t) = \prod (x_t | y_t)^{d\mu(t)}$$

$$\otimes \lambda_t x_t = \prod \lambda_t^{d\mu(t)} . \otimes x_t \qquad (\lambda_t \in \underline{C})$$

Moreover one can define the continuous tensor product of a family of operators U_t in H_t by

$$(\otimes U_t)(\otimes x_t) = \otimes U_t x_t .$$

c) Continuous products of measure spaces

The continuous product of a family of measure spaces (X_t, μ_t) is a measure space (X, μ) such that μ -almost every x in X has components x_t in X_t ; there exist sufficiently many functions f on X of the form

$$\otimes f_t \ : \ x \longmapsto \prod f_t(x_t)^{d\mu(t)}$$

where f_t is a function on X_t ; moreover we have

$$< \mu , \otimes f_t > \ = \prod < \mu_t, f_t >^{d\mu(t)}$$

and $L^2(X, \mu)$ is canonically isomorphic to the continuous tensor product of the spaces $L^2(X_t, \mu_t)$.

§ 8.2. Continuous products of complex numbers

We begin with a few examples.

Example 8.1. We denote by T an interval of $\underline{R}$ the lenght of which is a finite integer, by μ the Lebesgue measure on T, and by $\mathcal{E}$ the set of all continuous complex functions on T. For every f in $\mathcal{E}$ we set

$$\prod f = \begin{cases} 0 & \text{if } f(t) = 0 \quad \text{for some } t \\ \exp(\int \log f(t).d\mu(t)) & \text{if } f(t) \neq 0 \quad \forall t ; \end{cases}$$

here log means an arbitrary continuous determination of the logarithm ; the value of the exponential does not depend on the choice of the determination since $\mu(T)$ is an integer. Formulas (8.1), (8.2), (8.3) trivially hold.

Example 8.2. Here $T = \underline{R}^n$ with $n \geqslant 2$, μ is the Lebesgue measure and $\mathcal{E}$ the set $\mathcal{K}(T) + 1$ of all continuous complex functions on T which are equal to 1 outside some compact subset. The one-point compactification $T \cup \{\omega\}$ of T being simply connected there exists for every f in $\mathcal{E}$, $f(t) \neq 0 \;\; \forall t$, a unique continuous determination of $\log f$ such that $\log f(\omega) = 0$; we define $\prod f$ as in example 8.1 and we have the same properties.

Example 8.3. Here T is a Borel space with a positive Borel measure μ , $\mathcal{E}$ is the set of all positive measurable functions f such that the real function $\log f$ is integrable ; we define

$$\prod f = \exp\left(\int \log f(t).d\mu(t)\right)$$

and we have again the same properties ; but we have an additional property concerning partitions of T : for every partition of T into measurable subsets T_i, $\prod f$ is the (ordinary) product of the partial products $\exp\left(\int_{T_i} \log f(t).d\mu(t)\right)$.

Definition 8.1. Given a set T and a set $\mathcal{E}$ of complex functions on T which is closed under multiplication, conjugation and taking absolute values, a continuous product on $\mathcal{E}$ is a mapping $\prod$ of $\mathcal{E}$ into $\underline{C}$ satisfying (8.1),(8.2),(8.3).

Examples have been given above ; examples 8.1 and 8.2 lead us to consider continuous products of a topological type :

<u>Definition</u> 8.2. A continuous product in the sense of definition 8.1 is said to be <u>topological</u> if T is a locally compact topological space, $\mathcal{E}$ is the set of all nowhere vanishing functions in $\mathcal{K}(T) + 1$, and Π maps $\mathcal{E}$ into $\underline{C}^* = \underline{C} - \{0\}$ and is continuous for the topology on $\mathcal{E}$ defined in the following manner : for every compact subset K of T, the set $\mathcal{E}_K$ of all f in $\mathcal{E}$ satisfying $f(t) = 1$ for $t \in T - K$ is endowed with the topology of uniform convergence, and $\mathcal{E}$ is the inductive limit of the various $\mathcal{E}_K$.

Note that $\mathcal{E} = \Gamma_c(T, \underline{C}^*)$ (see definition in § 6.3). We denote by $M(T)$ the set of all real Radon measures on T. The next proposition gives an explicit description of the topological products.

<u>Proposition</u> 8.1. With the notations of definition 8.2 every f in $\Gamma_c(T, \underline{C})$ can be written $f = f'.e^{f''}$ where $f' = f/|f|$ and $f'' = \log|f|$; the topological continuous products on $\Gamma_c(T, \underline{C}^*)$ are exactly the mappings $f \longmapsto \chi'(f').e^{\chi''(f'')}$ where χ' is a continuous morphism $\Gamma_c(T, \underline{U}) \longrightarrow \underline{U}$ and χ'' a continuous morphism $\Gamma_c(T, \underline{R}) \longrightarrow \underline{R}$, i.e. a real Radon measure on T. Moreover $\Gamma_c(T, \underline{U})$ is isomorphic to the direct product $\Gamma_c(T, \underline{R})/\Gamma_c(T, \underline{Z}) \times [T, \underline{U}]_c$ where $[T, \underline{U}]_c$ is the group of all homotopy classes of the elements of $\Gamma_c(T, \underline{U})$. Finally the continuous morphisms $\Gamma_c(T, \underline{U}) \longrightarrow \underline{U}$ form a group isomorphic to the direct product $M_{\underline{Z}}(T) \times \widehat{[T, \underline{U}]_c}$ where $M_{\underline{Z}}(T)$ is the set of all real Radon measures on T taking integral values on $\Gamma_c(T, \underline{Z})$ and $\widehat{[T, \underline{U}]_c}$ is the group of all morphisms $[T, \underline{U}]_c \longrightarrow \underline{U}$.

We do not give the proof of this result, which can be found in [6], but we illustrate it with a few examples.

Example 8.4. We suppose that each connected component in T is open and closed, that each compact connected component is arcwise connected and simply connected, and that $T_1 \cup \{\omega\}$ is arcwise connected and simply connected where T_1 is the union of all non compact connected components and $T_1 \cup \{\omega\}$ is its one-point compactification. Then every element of $\Gamma_c(T,\underline{U})$ has a continuous logarithm, $[T,\underline{U}]_c$ is trivial, $M_{\underline{Z}}(T)$ is the set of all real Radon measures μ such that $\mu(C) \in \underline{Z}$ for every compact connected component C ; finally the topological continuous products are exactly the mappings

$$f \longmapsto \exp\left(\mu(\log f/|f|) + \nu(\log|f|)\right)$$

where $\mu \in M_{\underline{Z}}(T)$ and $\nu \in M(T)$; as in example 8.1 , the expression $\exp(\mu(\log f/|f|))$ is independent of the choice of the log since $\mu(C) \in \underline{Z}$ for every compact connected component C. This example includes examples 8.1 , 8.2 (there we had taken $\mu = \nu$) and also the case of ordinary products : T is discrete and $\mu = \nu$ has mass one at each point.

Example 8.5. We take $T = \underline{R}$; then $[T,\underline{U}]_c$ is isomorphic to $\underline{Z}$, $\widehat{[T,\underline{U}]_c}$ is isomorphic to $\underline{U}$ and $\Gamma_c(T,\underline{Z})$ is null. The isomorphism

$$F \ : \ \Gamma_c(T,\underline{U}) \longrightarrow \Gamma_c(T,\underline{R}) \times \underline{Z}$$

can be realized as follows : for every n in $\underline{Z}$ we set

$$\psi_n(t) = \begin{cases} e^{2\pi i n t} & \text{if} \quad 0 \leq t \leq 1 \\ 1 & \text{if not ;} \end{cases}$$

let f be an element of $\Gamma_c(T,\underline{U})$ and n its index with respect to the point 0; f/Ψ_n has index 0 and therefore can be written as $e^{2\pi ig}$ where $g \in \Gamma_c(T,\underline{R})$ and g is unique; then F(f) is the pair (g,n); we shall write

$$n = n(f)$$
$$g = (2\pi i)^{-1} \log(f/\Psi_{n(f)}) .$$

On the other hand $M_{\underline{Z}}(T)$ is equal to M(T). Thus the topological continuous products on $\mathcal{E}$ are exactly the mappings (we set $f' = f/|f|$) :

$$f \longmapsto \theta^{n(f)}.\exp\left[\mu(\log(f'/\Psi_{n(f)})) + \nu(\log|f|)\right.$$

where $\theta \in \underline{U}$, μ and $\nu \in M(T)$.

Remark 8.1. The additional property in example 8.3 could suggest a study of continuous products in another direction, of a Borel type : T is a Borel space with Borel structure $\mathcal{B}$, $\mathcal{E}$ is the set of all Borel mappings $T \longrightarrow \underline{C}^*$ taking only a finite number of values ; we define $\chi_{g,S}$ and $f^{(S)}$ as in § 6.1 and we assume, beside (8.1),(8.2),(8.3), the following axioms :

- $\Pi f \in \underline{C}^* \ \forall f \in \mathcal{E}$
- the mapping $z \longmapsto \Pi(\chi_{g,S})$ is continuous for every S in $\mathcal{B}$
- for every partition $S = \cup S_j$ with $S, S_j \in \mathcal{B}$, and every f in $\mathcal{E}$ we have

$$\sum_j |\Pi(f^{(S_j)}) - 1| < \infty$$

and

$$\prod_j \Pi(f^{(S_j)}) = \Pi(f^{(S)}) .$$

For instance we can take

$$\prod f = f(t_1)^{n_1} \ldots . f(t_k)^{n_k} . \exp\left(\int \log |f(t)| . d\mu(t)\right) \qquad (8.4)$$

where $t_1, \ldots t_k \in T$, $n_1, \ldots n_k \in \underline{Z}$ and μ is a finite real measure on T. Conversely every continuous product satisfying the above axioms is of that form. In fact for every S in $\mathcal{B}$, the mapping $z \longmapsto \prod(\chi_{g,S})$ is a continuous morphism $\varphi : \underline{C}^* \to \underline{C}^*$ satisfying $\varphi(\bar{z}) = \overline{\varphi(z)}$; writing $z = ab$ with $a = z/|z|$ and $b = |z|$ we see that φ has the form $\varphi(ab) = a^n b^u$ where $n \in \underline{Z}$ and $u \in \underline{R}$; we write $n = n(S)$, $u = u(S)$ and we have

$$\prod(\chi_{g,S}) = a^{n(S)}\, b^{u(S)} .$$

The third axiom implies that n and u are Borel measures ; n must be concentrated on a finite subset $t_1, \ldots t_k$ and we get (8.4) by setting $n_j = n(\{t_j\})$, $\mu = u$.

We thus see that the continuous products of the indicated type are not very interesting : a part from a finite product, $\prod f$ depends only on $|f|$.

§ 8.3. Continuous tensor products of Hilbert spaces

The definition of the tensor product of a finite family of Hilbert spaces relies heavily on the fact that the coefficientwise product of a finite family of positive definite matrices is again positive definite (see § 3.1) ; in fact such a tensor product $\overset{h}{\underset{i \in I}{\otimes}} H_i$ can be defined as follows : denote by Γ the product set

$\prod_{i \in I} H_i$, with elements $x = (x_i)_{i \in I}$; and by λ the following kernel on Γ : $\lambda(x,x') = \prod_{i \in I} (x_i | x'_i)$; λ is positive definite because it is the product of the kernels $(x,x') \longmapsto (x_i | x'_i)$ which are positive definite. Then $\overset{h}{\bigotimes_{i \in I}} H_i$ is nothing but the Hilbert space canonically associated with that kernel (see § 3.1) and $\otimes x_i$ is the canonical image of the element $\delta_x \in \underline{C}^{(\Gamma)}$.

The following proposition shows that the quoted property of finite componentwise products of matrices <u>does not hold</u> for continuous products.

<u>Proposition</u> 8.2. There exist continuous real nowhere vanishing functions a_{pq} on $T = [0,1]$, for $p,q = 1,2,3$, such that for every t in T the matrix $(a_{pq}(t))$ is positive definite but the matrix $(\exp \int_0^1 \log a_{pq}(t).dt)$ is not positive definite.

For any real numbers x_1, x_2, x_3 the matrix with coefficients $a_{pq} = \cos(x_p - x_q)$ is positive definite since for $c_1, c_2, c_3 \in \underline{C}$ we have

$$\sum_{pq=1}^{3} c_p \bar{c}_q \cos(x_p - x_q) = \sum c_p \bar{c}_q (\cos x_p . \cos x_q + \sin x_p . \sin x_q)$$

$$= |\Sigma c_p \cos x_p|^2 + |\Sigma c_p \sin x_p|^2 \geqslant 0.$$

Take a positive number ε and define functions f_1, f_2, f_3 on T by

$f_1(t) = 0 \quad \forall\, t$

$f_2(t) = -\varepsilon \quad \forall\, t$

$$f_3(t) = \begin{cases} 0 & \text{if } 0 \le t \le 1-2\sqrt{\varepsilon} \\ (\frac{\pi}{2} - \varepsilon - \varepsilon^{\frac{1}{\varepsilon}})(t - 1 + 2\sqrt{\varepsilon})/\sqrt{\varepsilon} & \text{if } 1 - 2\sqrt{\varepsilon} \le t \le 1 - \sqrt{\varepsilon} \\ \frac{\pi}{2} - \varepsilon - \varepsilon^{1/\varepsilon} & \text{if } 1 - \sqrt{\varepsilon} \le t \le 1 . \end{cases}$$

Set $a_{pq}(t) = \cos(f_p(t) - f_q(t)$; then $a_{pq}(t) \neq 0 \quad \forall\, t$ and the matrix having coefficients $\exp(\int \log a_{pq}(t).dt)$ is

$$M = \begin{pmatrix} 1 & c & c' \\ c & 1 & c'' \\ c' & c'' & 1 \end{pmatrix}$$

where

$$c = \exp(\int \log \cos(f_1(t)-f_2(t)).dt)$$
$$c' = \exp(\int \log \cos(f_3(t)-f_1(t)).dt)$$
$$c'' = \exp(\int \log \cos(f_3(t)-f_2(t)).dt) .$$

We have $c = \cos \varepsilon$ which tends to 1 when ε tends to 0. Now c' also tends to 1 since

$$2\sqrt{\varepsilon} \log \cos(\frac{\pi}{2} - \varepsilon) \le \int_0^1 \log \cos(f_3(t)-f_1(t)).dt \le 0 ;$$

finally c'' tends to 0 since

$$\int_0^1 \log \cos(f_3(t)-f_2(t)).dt \le \int_{1-\sqrt{\varepsilon}}^1 \log \cos(\frac{\pi}{2} - \varepsilon^{\frac{1}{\varepsilon}}).dt$$

$$= \sqrt{\varepsilon} \log \varepsilon^{\frac{1}{\varepsilon}} + \sqrt{\varepsilon} \log \frac{\sin \varepsilon^{\frac{1}{\varepsilon}}}{\varepsilon^{1/\varepsilon}}$$

For sufficiently small ε , M is not positive definite since the corresponding quadratic form takes on the vector $(-1,1,1)$ the value $3-2c-2c'+2c''$ which is not positive.

QED

We now proceed to the definition of the continuous tensor products of Hilbert spaces. Suppose we are given a set T, a family of Hilbert spaces $(H_t)_{t\in T}$, a subset Γ of the product set $\prod_{t\in T} H_t$ (its elements will be denoted $x = (x_t)$) and a kernel λ on Γ such that

$$\lambda(x,y) = \overline{\lambda(y,x)} \tag{8.5}$$

$$\lambda(x,x) \geq 0 \quad \forall x. \tag{8.6}$$

Intuitively $\lambda(x,y)$ is the continuous product of the complex numbers $(x_t|y_t)$ but in certain cases we cannot consider such a continuous product.

<u>Definition</u> 8.3. If the kernel λ is not positive definite, we say that the continuous tensor product does not exist ; if it is positive definite, we call <u>continuous tensor product</u> the Hilbert space canonically associated with it ; we denote it by $\overset{h(\Gamma,\lambda)}{\underset{t\in T}{\otimes}} H_t$; for every $x = (x_t)$ in Γ we denote by $\otimes x_t$ the canonical image of $\delta_x \in \underline{C}^{(\Gamma)}$. Elements of this type will be called <u>decomposable</u> ; they generate the continuous tensor product and satisfy

$$(\otimes x_t | \otimes y_t) = \lambda(x,y) \tag{8.7}$$

which implies

$$\| \otimes x_t - \otimes y_t \|^2 = \lambda(x,x) + \lambda(y,y) - 2 \operatorname{Re} \lambda(x,y) . \tag{8.8}$$

The continuous tensor products of unitary operators can now be defined in the following way : suppose we are given for each t a unitary operator U_t in H_t such that

(i) for every $x = (x_t)$ in Γ, the families $Ux = (U_t x_t)$ and $U^{-1}x = (U_t^{-1} x_t)$ belong to Γ

(ii) for every x and y in Γ we have $\lambda(Ux,Uy) = \lambda(x,y)$;

then for x and y in Γ we have

$$(\otimes x_t | \otimes y_t) = (\otimes U_t x_t | \otimes U_t y_t) \quad ;$$

since elements $\otimes x_t$ and $\otimes U_t x_t$ are total in $H = \overset{h}{\otimes}^{(\Gamma,\lambda)} H_t$ there exists a unique unitary operator $\otimes U_t$ in H such that

$$\otimes U_t . \otimes x_t = \otimes U_t x_t \qquad \forall\, x \in \Gamma .$$

Operators of this type will be called decomposable.

It is worth noticing that in many cases condition (ii) is automatically fulfilled.

Example 8.6. Suppose we have a continuous product Π on a set $\mathcal{E}$ as in definition 8.1, and that for any x and y in Γ, the function $t \longmapsto (x_t | y_t)$ belongs to $\mathcal{E}$; then the kernel $\lambda(x,y) = \Pi(x_. | y_.)$ satisfies (8.5) and (8.6) but it is not necessarily positive definite (see example 8.7) ; if it is positive definite we say that the continuous tensor product is defined by means of a continuous product of complex numbers ; (8.7) and (8.8) become

$$(\otimes x_t | \otimes y_t) = \Pi(x_. | y_.) \tag{8.9}$$

$$\|\otimes x_t - \otimes y_t\|^2 = \Pi(\|x_.\|^2) + \Pi(\|y_.\|^2) - 2\,\mathrm{Re}\,\Pi(x_. | y_.) \tag{8.10}$$

which yields a certain continuity of the mapping $(x_t) \longmapsto \otimes x_t$. Moreover condition (ii) above is automatically verified.

Example 8.7. We take $T, \mu, \mathcal{E}, P$ as in example 8.1 (here $T = [0,1]$) $H_t = \underline{C}^2$ with orthonormal basis a,b ; Γ = all continuous mappings of T into $\underline{C}^2$. Then the continuous tensor product <u>does not exist</u>. In fact we can define elements $x^{(j)}$ of Γ, j = 1,2,3, by

$$x_t^{(j)} = 2^{-\frac{1}{2}}(e^{if_j(t)} . a + e^{-if_j(t)} . b)$$

where f_1, f_2, f_3 are defined as in prop.8.2 ; then

$$\lambda(x^{(j)}, x^{(k)}) = \prod (x_{\cdot}^{(j)} | x_{\cdot}^{(k)})$$

$$= \exp \int_0^1 \log \cos(f_j(t) - f_k(t)).dt$$

and we know that the matrix with coefficients $\lambda(x^{(j)}, x^{(k)})$ is not positive definite.

<u>Example</u> 8.8. We keep the notations of § 6.3 , we set $\varphi_t(g) = e^{M(t,g)}$ and suppose φ_t is positive definite ; we set $H_t = H_{\varphi_t}$ and take for Γ the set of all families $x^{(f)}$, $f \in F^o$, where

$$x_t^{(f)} = \Lambda_{\varphi_t}(\delta_{f(t)}) ;$$

finally we define the kernel λ by

$$\lambda(x^{(f)}, x^{(f')}) = \varphi(f'^{-1} f) .$$

Then the continuous tensor product exists iff φ is positive definite ; if it exists it is canonically isomorphic to $H = H_\varphi$. This follows from the equality

$$(\otimes x_t^{(f)} | \otimes x_t^{(f')}) = \lambda(x^{(f)}, x^{(f')}) = \varphi(f'^{-1} f)$$

$$= (\Lambda_\varphi(\delta_f) | \Lambda_\varphi(\delta_{f'}))$$

and from the fact that elements $\otimes x_t^{(f)}$ and $\Lambda_\varphi(\delta_f)$ are total respectively in $\overset{h(\Gamma,\lambda)}{\otimes} H_t$ and in H . Moreover it is easily verified that this isomorphism carries $\otimes U_{\varphi_t}(f(t))$ into $U_\varphi(f)$ for every f in F^0.

Now we would like to write

$$\varphi = \otimes \varphi_t$$

i.e.

$$\varphi(f) = \prod \varphi_t(f(t))^{d\mu(t)} \qquad \forall\ f \in F^0 \tag{8.11}$$

which would make more natural the equalities $H_\varphi = \otimes H_t$ and $U_\varphi = \otimes U_t$; unfortunately (8.11) does not always make sense ; it does make sense (and is actually true) if T satisfies the conditions of example 8.4 as well as μ ; or if T and μ are arbitrary but M is real (for instance Gaussian) ; in fact in both cases $M(t,f(t))$ can be considered as the logarithm of its exponential. We emphasize the following interesting feature of this example : the function $\otimes \varphi_t$, even when it is truly defined, and when each φ_t is positive definite, is not necessarily positive definite ; one has to assume more than the positive definiteness of the φ_t's , for instance that they are infinitely divisible.

Example 8.9. (more particular than the preceding one, but (T,μ) can be an arbitrary finite measure space). We take for G the group with two elements e, e' and for M the function

$$M(t,g) = M(g) = \begin{cases} 0 & \text{if } g = e \\ -1 & \text{if } g = e' ; \end{cases}$$

prop.6.1 shows that φ is positive definite since M is conditionally positive definite. Setting $\varepsilon = \Lambda_{\varphi_t}(\delta_e)$, $\varepsilon' = \Lambda_{\varphi_t}(\delta_{e'})$

we have

$$\|\varepsilon\| = \|\varepsilon'\| = 1 \ , \quad (\varepsilon \mid \varepsilon') = 1/e$$

therefore H_t is two-dimensional with basis $(\varepsilon, \varepsilon')$; Γ is the set of all measurable mappings $x = (x_t) : T \longrightarrow \underline{C}^2$ such that $x_t = \varepsilon$ or ε' for every t.

We thus get the following example of an existing continuous tensor product of Hilbert spaces (compare with example 8.7 !) : (T, μ) is an arbitrary finite measure space, H_t is equal to $\underline{C}^2$ where we have chosen a basis ε, ε' satisfying $\|\varepsilon\| = \|\varepsilon'\| = 1$ and $(\varepsilon \mid \varepsilon') = 1/e$; Γ is the set of all measurable mappings x of T into $\underline{C}^2$ suchthat $x_t = \varepsilon$ or ε' $\forall$ t ; finally the kernel λ is

$$\lambda(x,x') = \exp \int \log (x_t \mid x'_t).d\mu(t) \ .$$

§ 8.4. Continuous tensor products of symmetric Hilbert spaces

Proposition 2.3 can be generalized as follows :

Proposition 8.3. Let us consider a Hilbert integral $K = \int_T^{\oplus} K_t.d\mu(t)$ where T is a Borel space and μ a positive measure ; set $H_t = SK_t$; let Γ be the set of all families $x^{(k,a)}$, $k \in \underline{C}$, $a = (a_t) \in K$, where $x_t^{(k,a)} = k.\mathrm{EXP}\ a_t$; define the kernel λ by

$$\lambda(x^{(k,a)}, x^{(k',a')}) = k\,k'\,e^{(a|a')}. \tag{8.11'}$$

Then the continuous tensor product $\overset{h}{\otimes}{}^{(\Gamma,\lambda)} H_t$ exists and there is a unique isomorphism Ω of it onto SK carrying every vector $\otimes\ k.\mathrm{EXP}\ a_t$ into $k.\mathrm{EXP}\ a$.

We first note that (8.11') makes sense since $k.\mathrm{EXP}\, a_t = k'.\mathrm{EXP}\, a'_t$ implies $k = k'$ and $a_t = a'_t$; the continuous tensor product exists since the kernel λ is trivially positive definite ; moreover

$$
\begin{aligned}
(\otimes\, k.\mathrm{EXP}\, a_t \mid \otimes\, k'.\mathrm{EXP}\, a'_t) &= \lambda\,(x^{(k,a)}, x^{(k',a')}) \\
&= k\,\bar{k'}\; e^{(a|a')} \\
&= (k.\mathrm{EXP}\, a \mid k'.\mathrm{EXP}\, a')
\end{aligned}
$$

and the last assertion follows from the fact that elements of the form $\otimes\, k.\mathrm{EXP}\, a_t$ and $k.\mathrm{EXP}\, a$ are total respectively in the spaces $\overset{h}{\otimes}{}^{(\Gamma,\lambda)} H_t$ and SK.

Remark 8.2. In certain cases the continuous tensor product can be defined by means of a continuous product of complex numbers : take for instance $T = [0,1]$, μ = Lebesgue measure, $K = L^2(T,\mu)$, Γ' = set of all families $x^{(k,a)}$ where $a \in \mathscr{C}(T)$, the set of all continuous functions on T ; then we have

$$
\begin{aligned}
\lambda\,(x^{(k,a)}, x^{(k',a')}) &= k\,\bar{k'}\,\exp\Big(\int a_t\,\bar{a}'_t\, d\mu\,(t)\Big) \\
&= k\,\bar{k'}\,\prod\,(\exp(a_t\,\bar{a}'_t))^{d\mu(t)} \\
&= \prod\,(x_t^{(k,a)} \mid x_t^{(k',a')})^{d\mu(t)} \; ;
\end{aligned}
$$

moreover (8.10) shows that elements $\otimes\, x_t$ with $x \in \Gamma'$ are total in the continuous tensor product, so that we have

$$\overset{h}{\otimes}{}^{(\Gamma,\lambda)}\, SK_t = \overset{h}{\otimes}{}^{(\Gamma',\lambda)}\, SK_t \,.$$

Onthe other hand the subspace $E = \mathscr{C}(T)$, dense in K, can be considered as a continuous sum of spaces E_t equal to $\underline{C}$; and we see that, to some extent, the Fock representation of a continuous sum of spaces E_t is the continuous tensor product of the Fock representations of the various E_t's (see App.F).

We can now reinterpret th.5.1 as follows : we keep the notations of th.5.1 and suppose that H is separable and that Θ is the Boolean algebra of all equivalence classes of measurable sets for a measure μ on a set T ; then K is separable too and there exists a desintegration $K = \int_T K_t.d\mu(t)$ such that for every θ in Θ , $K_\theta = \int_\theta K_t.d\mu(t)$; composing the isomorphisms $\Phi_{\mathbb{1}}$ of th.5.1 and Ω^{-1} of prop.8.3 we obtain an isomorphism of H onto $\overset{h}{\otimes}{}^{(\Gamma,\lambda)} H_t$ which carries H^1 onto the set of all elements $\otimes x_t^{(k,a)}$. We have thus proved the following

<u>Theorem</u> 8.1. With the notations of theorem 5.1 we suppose that H is separable and that Θ is the Boolean algebra of all equivalence classes of measurable subsets for a measure μ on a set T . Then there exists a decomposition of H in a continuous tensor product $\overset{h}{\underset{t\in T}{\otimes}}{}^{(\Gamma,\lambda)} H_t$ such that the factorizable vectors in H are exactly the decomposable vectors in the continuous tensor product. Moreover the spaces H_t are of the form SK_t , the components x_t of the decomposable vectors $\otimes x_t$ are of the form $k.\mathrm{EXP}\, a$, $k \in \underline{C}^*$, $a \in K_t$.

<u>Remark</u> 8.3. Many continuous tensor products of Hilbert spaces

give rise to Boolean algebras of tensor decompositions, for instance examples 8.8 and 8.9 ; theorem 8.1 seems to prove that such a continuous tensor product must be of the type of prop.8.3 ; howver this is not true for example 8.9 since there $\dim H_t = 2$. On the other hand it would be interesting to complete th.8.1 by saying that the factorizable unitary operators in H are exactly the decomposable unitary operators in the continuous tensor product ; to achieve this we should identify $U_{A,b,c}$ with the operator $\otimes U_{A_t,b_t,c_t}$ where $A = \int^{\oplus} A_t.d\mu(t)$, $b = \int^{\oplus} b_t.d\mu(t)$, $c = \prod c_t^{d\mu(t)}$; we can do it for instance by assuming that the K_t's are real Hilbert spaces and replacing Γ in prop.8.3 by $\tilde{\Gamma} = \{x^{(k,a)}\}$ where $x_t^{(k,a)} = k_t.\mathrm{EXP}\, a_t$, $a \in K$ and k is a function of the form e^h with $h \in L^1(T,\mu\, ;\underline{R})$.

§ 8.5. Continuous products of measure spaces

We begin with a few examples.

Example 8.10. We denote by Z and T two standard Borel spaces, by $\mathcal{a}$ the set of all Borel subsets in T, by π a finite positive Borel measure on Z , by p a Borel mapping of Z onto T , by ν the image of π under p ; we set $Z_t = p^{-1}(\{t\})$ $\forall t \in T$; there exists a desintegration $\pi = \int \pi_t.d\nu(t)$ where π_t is a normalized measure carried by Z_t . We set (see chapter 1)

$$X = SZ \quad , \quad \alpha = 0 \text{ element of } X$$

$$\mu = S\pi$$

and similarly

$$X_t = SZ_t \ , \quad \alpha_t = 0 \text{ element of } X_t$$

$$\mu_t = S\pi_t \ ;$$

then we would like to consider the measure space (X, μ) as the continuous product of the measure spaces (X_t, μ_t) ; this is justified by the following considerations :

a) If we describe X in the N-picture (see §1.1), every element N of X is characterized by a family of components $N_t \in X_t$ where N_t is the restriction of N to Z_t . If f is a Borel function on Z and $f_t = f|Z_t$, we have

$$\overset{\infty}{f}(N) = \prod_{z\in Z} f(z)^{N(z)} = \prod_{t\in T} \prod_{z\in Z_t} f_t(z)^{N_t(z)}$$

$$= \prod_{t\in T} \overset{\infty}{f_t}(N_t)$$

and we can write $\overset{\infty}{f} = \otimes \overset{\infty}{f_t}$; if moreover f is π -integrable we have by formula (1.6)

$$< \mu , \overset{\infty}{f} > = \exp < \pi , f > = \exp \int < \pi_t, f_t > .dv(t)$$

or formally

$$< \mu , \overset{\infty}{f} > = \prod \exp < \pi_t, f_t >^{dv(t)}$$

$$= \prod < \mu_t, \overset{\infty}{f_t} >^{dv(t)} ; \qquad (8.12)$$

this formula becomes rigorous if f is real (see example 8.3). Moreover by §1.4, property (v) we know that there are sufficiently many of these functions : the functions $\overset{\infty}{f}$ with $f \in L^2(Z, \pi)$ are total in $L^2(X, \mu)$ and therefore in $L^1(X, \mu)$.

b) It follows that there is a unique isomorphism

$$\overset{h}{\otimes}{}^{(\Gamma,\lambda)} L^2(X_t, \mu_t) \longrightarrow L^2(X, \mu)$$

carrying every element $\overset{\infty}{\otimes} f_t$ into $\overset{\infty}{f}$, where Γ is the set of all families $x^{(f)}$, $f \in L^2(Z, \pi)$, $x_t^{(f)} = \overset{\infty}{f_t}$ and λ is defined by

$$\lambda(x^{(f)}, x^{(g)}) = \exp < \pi , f \bar{g} > ;$$

note that this is formally (but rigorously for real f and g) equal to $\prod (x_t^{(f)} | x_t^{(g)})^{d\nu(t)}$.

c) For every $A \in \mathcal{A}$ we set (see § 1.5)

$$Z_A = p^{-1}(A) \quad , \quad \pi_A = \pi | Z_A$$

$$X_A = S(Z_A) \quad , \quad \alpha_A = 0 \text{ element of } X_A$$

$$\mu_A = S(\pi_A) ;$$

let Θ denote the Boolean algebra of all ν-equivalence classes of elements of $\mathcal{A}$; if A is ν-negligible, Z_A is π-negligible, π_A is null, μ_A is concentrated on α_A ; therefore if A and A' are ν-equivalent, the measure spaces (X_A, μ_A) and $(X_{A'}, \mu_{A'})$ are canonically isomorphic ; we write (X_θ , μ_θ) instead of (X_A, μ_A) where θ is the equivalence class of A . For every partition $\mathcal{F} : \theta = \vee \theta_i$ in Θ we have by § 1.4, property (viii), a measure preserving isomorphism (see definition in App. D.2)

$$M_{\mathcal{F}} : \prod^{(\alpha_{\theta_i})} (X_{\theta_i}, \mu_{\theta_i}) \longrightarrow (X_\theta , \mu_\theta)$$

which satisfies associativity and commutativity conditions quite similar to conditions (iii) and (iv) in definition 5.1.

Remark 8.4. Unfortunately μ_t is not the image of μ (even after normalization of μ_t and μ) under the mapping $P : N \longmapsto N_t$ as is the case for ordinary products of measure spaces. In fact, as easily checked, the function $\overset{\circ}{1}$ on X_t is not μ_t-negligible while $\overset{\circ}{1} \circ P$ is μ-negligible if ν is non atomic.

Example 8.11. We keep the notations of example 8.8 : T is a compact topological space, ν a positive measure on T , G a separable commutative locally compact group, F the set of all Borel mappings $T \longrightarrow G$ taking only a finite number of values, ε is the neutral element of F , M is a continuous function on $T \times G$ identically zero on $T \times \{e\}$; we set

$$\varphi_t(g) = \exp(M(t,g)) \qquad \forall\, t \in T\,,\ g \in G$$

$$\varphi(f) = \exp\left[\int M(t,f(t)).d\nu(t)\right] \quad \forall\, f \in F\ ;$$

we suppose φ and φ_t positive definite. By App.D.2 we can write $H_\varphi = L^2(X,\mu)$ where (X,μ) is some measure space, $\xi_\varphi = 1$, $U_\varphi(f)$ = multiplication operator by a function $\psi(f)$; and similarly $H_{\varphi_t} = L^2(X_t,\mu_t)$, $\xi_{\varphi_t} = 1$, $U_{\varphi_t}(g)$ = multiplication by $\psi_t(g)$.

We would like to consider (X,μ) as the continuous product of the various (X_t,μ_t) and this is justified by properties analogous to those of example 8.10, except that here we have no mappings $X \longrightarrow X_t$.

a) For every f in F we have elements $\Lambda_\varphi(\delta_f)$ in H_φ and $\Lambda_{\varphi_t}(\delta_{f(t)})$ in H_{φ_t} ; by the above identification these elements become functions $\alpha^{(f)} \in L^2(X, \mu)$ and $\alpha_t^{(f)} \in L^2(X_t, \mu_t)$; we have

$$\begin{aligned} < \mu, \alpha^{(f)} > &= (\alpha^{(f)} | \alpha^{(\varepsilon)}) = (\Lambda_\varphi(\delta_f) | \Lambda_\varphi(\delta_\varepsilon)) \\ &= \varphi(f) = \exp\left[\int M(t,f(t)).d\nu(t)\right] \end{aligned}$$

and then formally (but rigorously if M is real)

$$\begin{aligned} &= \prod \exp M(t,f(t))^{d\nu(t)} \\ &= \prod \varphi_t(f(t))^{d\nu(t)} \\ &= \prod < \mu_t, \alpha_t^{(f)} >^{d\nu(t)} \ ; \end{aligned}$$

writing $\alpha^{(f)} = \otimes \alpha_t^{(f)}$ we get a formula similar to (8.12) ; moreover the functions $\alpha^{(f)}$ are total in $L^2(X, \mu)$ and hence in $L^1(X, \mu)$.

b) Property analogous to property b) in example 8.10 has been proved in example 8.8.

c) For every Borel subset A of T we can define similarly φ_A , H_A , U_A , X_A , μ_A , which depend only on the ν-equivalence class θ of A and which we write φ_θ , H_θ , U_θ , X_θ , μ_θ . For every partition $\mathcal{F} : \theta = \vee\, \theta_i$ we have an isomorphism

$$\overset{h}{\otimes}{}^{(1)}\, L^2(X_{\theta_i}, \mu_{\theta_i}) \longrightarrow L^2(X_\theta, \mu_\theta)$$

carrying $\otimes 1$ into 1 and $\otimes U_{\theta_i}(g_i)$ into $U_\theta(g)$ for every g in $F^{(\theta)}$ with $g_i = g|\theta_i$; on the other hand we have an isomorphism

$$L^2(\prod X_{\theta_i}, \otimes \mu_{\theta_i}) \longrightarrow \overset{h}{\otimes}{}^{(1)} L^2(X_{\theta_i}, \mu_{\theta_i})$$

carrying 1 into $\otimes 1$; whence an isomorphism

$$L^2(\prod X_{\theta_i}, \otimes \mu_{\theta_i}) \longrightarrow L^2(X_\theta, \mu_\theta) \quad ;$$

its inverse carries all multiplication operators into multiplication operators ; then our isomorphism comes from an isomorphism

$$M_{\mathcal{F}} : (\prod X_{\theta_i}, \otimes \mu_{\theta_i}) \longrightarrow (X_\theta, \mu_\theta) \ ;$$

these isomorphisms $M_{\mathcal{F}}$ satisfy conditions similar to those in example 8.10.

In a similar manner we can take $T = \underline{R}^n$, $G = \underline{R}$ and replace F by $\mathcal{D}(\underline{R}^n;\underline{R})$; by Bochner-Minlos theorem φ is the Fourier transform of a measure μ on $\mathcal{D}'$, and by the classical Bochner theorem φ_t is the Fourier transform of a measure μ_t on $\underline{R}_t = \underline{R}$. We see that the measure space $(\mathcal{D}', \mu)$ can be considered as the continuous product of the measure spaces $(\underline{R}_t, \mu_t)$, this being justified by properties analogous to a),b),c).

<u>Notion of continuous product of measure spaces</u>

The two preceding examples lead us to consider a measure space (X, μ) as the continuous product of a family (X_t, μ_t) if, roughly speaking

a) there exist sufficiently many functions f in $L^1(X, \mu)$ corresponding to some families (f_t) with $f_t \in L^1(X_t, \mu_t)$ and (formally) $\langle \mu, f \rangle = \prod \langle \mu_t, f_t \rangle^{d\nu(t)}$.

b) there exists an isomorphism of $L^2(X,\mu)$ onto some continuous tensor product $\otimes L^2(X_t,\mu_t)$

c) for every Borel subset A of T we have a measure space (X_A,μ_A) with the following property : for every partition $A = \cup A_i$, (X_A,μ_A) is some restricted product of the (X_{A_i},μ_{A_i}).

However we shall not try here to give a precise definition, but only to discuss the following notion (rather similar to that of definition 5.1) which we are lead to by property c) above.

Definition 8.4. Let (X,μ) be a standard measure space and Y a Borel subset of X such that $\mu(Y) = 1$ (we emphasize that in general we have $\mu(X) > 1$) ; we call Boolean algebra of product decompositions (BAPD) of (X,μ,Y) a family of four objects Θ, $((X_\theta,\mu_\theta))_{\theta\in\Theta}$, $(Y_\theta)_{\theta\in\Theta}$, $(M_{\mathcal{F}})_{\mathcal{F}\in\widetilde{\Theta}}$ where

- Θ is an abstract complete Boolean algebra
- for every θ in Θ, (X_θ,μ_θ) is a measure space and Y_θ a Borel subset of X_θ with $\mu_\theta(Y_\theta) = 1$
- $\widetilde{\Theta}$ is the set of all partitions in Θ
- for every $\mathcal{F} = (\theta_i)_{i\in I} \in \widetilde{\Theta}$, $M_{\mathcal{F}}$ is an isomorphism of $\prod^{(Y_{\theta_i})}(X_{\theta_i},\mu_{\theta_i})$ onto $(X_{\vee\theta_i},\mu_{\vee\theta_i})$ carrying $\prod Y_{\theta_i}$ onto $Y_{\vee\theta_i}$

satisfying the following conditions :

(i) $X_{\mathbb{1}} = X$, $\mu_{\mathbb{1}} = \mu$, $Y_{\mathbb{1}} = Y$

(ii) $X_{\mathbb{0}}$ is reduced to a single point with measure 1

(iii) and (iv) : associativity and commutativity conditions similar to conditions (iii) and (iv) in definition 5.1.

Example 8.10 provides an example of BAPD with Y_θ reduced to a single point α_θ ; and example 8.11 another with $Y_\theta = X_\theta$; we have no examples where Y_θ is a strict subset of X_θ distinct from one point. The following theorem shows that when Y_θ is reduced to a single point example 8.10 is general under some additional assumptions.

Definition 8.5. Given a BAPD as in definition 8.4, a complex Borel function f on X is called factorizable if for every finite partition of $\mathbb{1}$: $\mathcal{F} = (\theta_i)$, there exist Borel functions f_i on X_{θ_i} such that $f \circ M_f = \otimes f_i$.

In example 8.10 the factorizable functions of modulus 1 are total by § 1.4, prop.(v) ; in example 8.11 this holds trivially.

Theorem 8.2. We consider a Boolean algebra of product decompositions as in definition 8.4 ; we assume that Θ is non atomic, that Y_θ is reduced to a single point α_θ , that the factorizable functions of modulus 1 on X are total in $L^2(X,\mu)$, and that the $M_{\mathcal{F}}$ are measure preserving. Then there exist two standard measure spaces (Z,π) , (T,ν), a Borel mapping p of Z onto T such that $p(\pi) = \nu$, a σ-morphism $\theta \longmapsto T_\theta$ of Θ onto the ν-equivalence classes of Borel subsets of T and for every θ in Θ a measure preserving isomorphism $\Omega_\theta : (X_\theta , \mu_\theta) \longrightarrow (S(p^{-1}(T)) , S(\pi \mid p^{-1}(T)))$ carrying α_θ into the 0 element of $S(p^{-1}(T_\theta))$ and such that the isomorphisms $M_{\mathcal{F}}$ become the canonical isomorphisms described in example 8.10.

Proof.

a) We set $H = L^2(X,\mu)$, $\omega = \delta_\alpha \in H$ and similarly $H_\theta = L^2(X_\theta, \mu_\theta)$, $\omega_\theta = \delta_{\alpha_\theta}$; by App.D.3 each isomorphism $M_{\widetilde{\mathcal{F}}}$ determines an isomorphism

$$\Lambda_{\widetilde{\mathcal{F}}} : \overset{h\ (\omega_{\theta_i})}{\otimes} H_{\theta_i} \longrightarrow H_\theta ;$$

we thus get a BATD , the conditions of definition 5.1 being easily verified. Let G be the set of all factorizable functions on X of modulus 1 satisfying $f(\alpha) = 1$; G is total in H by assumption and included in H^o (see notations of th.5.1) ; hence we can apply theorems 5.1 and 5.2 : there exist a Hilbert space K , a family of subspaces K_θ and isomorphisms $\Phi_\theta : H_\theta \longrightarrow SK_\theta$ such that we have (writing $\Phi = \Phi_{\mathbb{1}}$) :

$$\Phi_\theta\ (\omega_\theta) = \text{EXP } 0$$

$$\Phi\ (H^o) = \text{set of all vectors EXP } a\ ,\ a \in K\ .$$

Φ transforms every factorizable unitary operator in H into an operator of the form $U_{A,b,c}$, where moreover A preserves all K_θ .

b) The set G is a (multiplicative) commutative group ; for every f in G , the multiplication operator T_f in H by f is a factorizable unitary operator ; $f \longmapsto T_f$ is a unitary representation of G ; moreover T_f preserves ω . Under Φ , T_f becomes an operator $U_{A(f),b(f),c(f)}$ which preserves EXP 0 ; thus $b(f) = 0$ and $c(f) = 1$ and we have

$$\Phi . T_f . \Phi^{-1} = U_{A(f),0,1}$$

where A is a unitary representation of G in K and A(f) preserves all K_θ .

c) The function 1 on X belongs to G, therefore $\Phi(1)$ is of the form EXP a , $a \in K$; 1 is cyclic for the operators T_f , $f \in G$, so that EXP a is cyclic for the operators $U_{A(f),0,1}$. We claim that a must be cyclic for the operators A(f). Suppose it is not cyclic ; we can write $K = K' \oplus K''$, K" distinct from 0, stable under the A(f) and orthogonal to all A(f).a ; then

$$SK = SK' \otimes SK''$$

$$U_{A(f),0,1} = U_{A(f)|K',0,1} \otimes U_{A(f)|K'',0,1}$$

$$\text{EXP } a = \text{EXP } a \otimes \text{EXP } 0$$

$$U_{A(f),0,1}(\text{EXP } a) = \text{EXP } (A(f).a) \otimes \text{EXP } 0 \ ;$$

since K" is not 0 , dim SK" is infinite, there exists u in SK" which is orthogonal to EXP 0 ; for every v in SK' , $v \otimes u$ is orthogonal to $U_{A(f),0,1}$ (EXP a) - a contradiction.

d) By App.D,2 there exist a standard measure space (Z,π) and an isomorphism $\psi : K \longrightarrow L^2(Z,\pi)$ carrying a into the function 1 and every A(f) into the multiplication operator by some function φ_f of modulus 1 . By prop.2.6 there is an isomorphism $\psi' : SK \longrightarrow L^2(SZ,S\pi)$ carrying $U_{A(f),0,1}$ into the multiplication operator by $\overset{\circ\circ}{\varphi}_f$, EXP 0 into δ_γ where γ is the 0 element of SZ , and EXP a into $\overset{\circ\circ}{1} = 1$.

e) The composition $\psi' \circ \Phi$ is an isomorphism $H \longrightarrow L^2(SZ,S\pi)$ carrying every T_f into $T_{\overset{\circ\circ}{\varphi}_f}$, $\omega = \delta_\alpha$ into δ_γ and 1 into 1 .

Since the T_f's have a cyclic vector they generate the von Neumann algebra $\mathcal{a}(X,\mu)$ of all diagonalisable operators (see App.D.2) ; $\psi' \circ \phi$ carries $\mathcal{a}(X,\mu)$ onto $\mathcal{a}(SZ,S\pi)$ and therefore it comes from an isomorphism $\Omega : (X,\mu) \longrightarrow (SZ,S\pi)$; a priori $\Omega(\mu)$ is only equivalent to $S\pi$, but since $\psi' \circ \phi$ carries 1 into 1, $\Omega(\mu)$ is equal to $S\pi$; moreover $\Omega(\alpha) = \gamma$.

f) ψ carries each K_θ onto a subspace of $L^2(Z,\pi)$ which is invariant under all T_{φ_f}, hence under all diagonalisable operators ; therefore this subspace is of the form $L^2(Z_\theta, \pi | Z_\theta)$ where Z_θ is a Borel subset of Z ; we thus have a σ-morphism $\theta \longmapsto Z_\theta$ of Θ into the π-equivalence classes of Borel subsets of Z. It is known (see [37], § 0) that there exist a standard measure space (T,ν), a Borel mapping p of Z onto T and a σ-morphism $\theta \longmapsto T_\theta$ of Θ onto the ν-equivalence classes of Borel subsets of T such that $p(\pi) = \nu$ and for every θ, Z_θ is the π-equivalence class of the inverse image under p of the elements of T_θ.

g) In the same manner as before we can construct isomorphisms $\Omega_\theta : (X_\theta, \mu_\theta) \longmapsto (SZ_\theta, S\pi_\theta)$ with $\pi_\theta = \pi | Z_\theta$; finally it is easy to see that the isomorphisms Ω_θ have the properties listed in the statement .

Remark 8.5. There is a close connection between the preceding result and the work of Feldman [11]; his definition of factored probability spaces (see def.1.3 of the present lecture) is another attempt toward the definition of continuous products of measure spaces. A BAPD is a particular case of factored measure space ; in fact by Loomis theorem Θ is a quotient of a σ-algebra $\mathcal{A}$ of subsets of a set T ; for every $A \in \mathcal{A}$, X is decomposed into a product $X_\theta \times X_{\theta'}$, where θ is the equivalence class of A ; we can define $\mathcal{B}_A$ as the inverse image of the σ-algebra of X_θ under the projection $X \longrightarrow X_\theta$, and we get a factored measure space (condition (iv) in def.1.3 follows from prop.D.1). Then theorem 8.2 gives an explicit description of these particular factored measure spaces when Y_θ is reduced to a single point ; and prop.1.1 shows that they are generated by Poissonian decomposable processes.

Appendix A. INFINITE TENSOR PRODUCTS OF HILBERT SPACES

(See [14], Part II)

§ A.1. Definitions

Suppose we are given a family of (real or complex) Hilbert spaces $(H_i)_{i \in I}$ and for each i, a unit vector a_i in H_i; for each finite subset J of I we construct the Hilbert tensor product $\overset{h}{\underset{i\in J}{\otimes}} H_i$. We make this family of Hilbert spaces into an inductive system inthe following way : suppose $J \subset K$ are finite subsets of I ; we can write

$$\overset{h}{\underset{i\in K}{\otimes}} H_i = \left(\overset{h}{\underset{i\in J}{\otimes}} H_i\right) \overset{h}{\otimes} \left(\overset{h}{\underset{i\in K-J}{\otimes}} H_i\right)$$

and define an isometric mapping

$$\overset{h}{\underset{i\in J}{\otimes}} H_i \longrightarrow \overset{h}{\underset{i\in K}{\otimes}} H_i$$

$$x \longmapsto x \otimes \left(\underset{i\in K-J}{\otimes} a_i\right) .$$

The inductive limit of this inductive system is denoted by $\overset{h}{\underset{i\in I}{\otimes}}{}^{a} H_i$ or $\overset{h\,(a_i)}{\otimes} H_i$; roughly speaking it is the completion of the union of all finite tensor products $\overset{h}{\underset{i\in J}{\otimes}} H_i$. We now denote it by H , and by F_J the canonical injection of the finite tensor product $\overset{h}{\underset{i\in J}{\otimes}} H_i$ into H.

Consider a family of vectors $x_i \in H_i$ with $x_i = a_i$ for almost every i , say $x_i = a_i$ if i does not belong to some finite subset J ; form $\bigotimes_{i \in J} x_i \in \overset{h}{\bigotimes}_{i \in J} H_i$; its canonical image in H is denoted by $\bigotimes_{i \in I} x_i$ or $\otimes x_i$. Elements of that form are called <u>elementary decomposable vectors</u> and generate H ; their scalar product is given by

$$(\otimes x_i | \otimes y_i) = \prod (x_i | y_i) .$$

We now indicate how to get an orthonormal basis of H ; choose for each i an orthonormal basis $(a_{i,u})_{u \in U_i}$ of H_i where U_i is a set containing an element v_i such that $a_{i,v_i} = a_i$; choose for each i an element u_i in U_i such that $u_i = v_i$ a.e. and form the vector $\otimes a_{i,u_i}$; then by varying the family (u_i) we get an orthonormal basis of H. It follows that H is separable if the H_i's do and if I is countable.

§ A.2. Associativity and commutativity

We begin with associativity : for every partition $I = \bigcup_{u \in U} I_u$ there exists a unique isomorphism G of $\overset{h}{\bigotimes}{}^{a}_{i \in I} H_i$ onto the space $\overset{h}{\bigotimes}{}^{b}_{u \in U} (\overset{h}{\bigotimes}{}^{a}_{i \in I_u} H_i)$ carrying each element $\bigotimes_{i \in I} x_i$ into the element $\bigotimes_{u \in U} (\bigotimes_{i \in I_u} x_i)$; here we have set $b_u = \bigotimes_{i \in I_u} a_i$.

The commutativity reads as follows : take a permutation s of I ; there exists a unique isomorphism of $\overset{h}{\bigotimes}{}^{a}_{i \in I} H_i$ onto the space

$\overset{h\,(a_{s(i)})}{\bigotimes_{i\in I}} H_{s(i)}$ carrying every element $\bigotimes_{i\in I} x_i$ into $\bigotimes_{i\in I} x_{s(i)}$.

§ A.3. Decomposable vectors

Consider a family (x_i) satisfying $x_i \in H_i$ and

$$\sum | \| x_i \| - 1 | < \infty \tag{A.1}$$

$$\sum |(x_i \; a_i) - 1| < \infty ; \tag{A.2}$$

then the infinite product $\prod \| x_i \|$ exists, and it is null iff some of the x_i's is null ; the family of the vectors $L_J(\bigotimes_{i\in J} x_i)$ has a limit in H, the norm of which is $\prod \| x_i \|$; this limit is denoted by $\otimes x_i$; such a vector will be called <u>decomposable</u> . For any two decomposable vectors we have

$$\sum |(x_i \; y_i) - 1| < \infty \tag{A.3}$$

$$(\otimes x_i | \otimes y_i) = \prod (x_i | y_i). \tag{A.4}$$

Moreover for any partition $I = \bigcup_{u\in U} I_u$, the isomorphism G of § A.2 carries $\bigotimes_{i\in I} x_i$ into $\bigotimes_{u\in U} (\bigotimes_{i\in I_u} x_i)$.

<u>Proposition</u> A.1. Let x be a non zero element of H with the following property : for every <u>finite</u> partition $I = \bigcup_{u\in U} I_u$ there exist elements $x_{(u)} \in \overset{h}{\bigotimes_{i\in I_u}}{}^{a} H_i$ such that $G(x) = \bigotimes_{u\in U} x_{(u)}$. Then x is decomposable.

(See [2], lemma 3.2)

__Proof__. Since x is not null there exists an element $b = \otimes b_i$ with $b_i = a_i$ a.e. and $(x|b) \neq 0$; let $K = \{i_1, \dots i_n\}$ be the set of all i such that $b_i \neq a_i$; using the finite partition

$$I = \{i_1\} \cup \dots \cup \{i_n\} \cup (I-K)$$

we can write

$$x = (\bigotimes_{i\in K} x_i) \otimes x'$$

where $x_i \in H_i$, $x' \in \overset{h}{\bigotimes}{}^{a}_{i\in I-K} H_i$ and

$$b = (\bigotimes_{i\in K} b_i) \otimes (\bigotimes_{i\in I-K} a_i)$$

$$(x|b) = \prod_{i\in K} (x_i | b_i) \times (x' | \bigotimes_{i\in I-K} a_i) \quad ;$$

this implies $(x' | \bigotimes_{i\in I-K} a_i) \neq 0$; since it is enough to prove that x' is decomposable, we are led to prove the proposition when $(x|a) \neq 0$ (we set $a = \otimes a_i$).

We can suppose $\|x\| = 1$ and $(x|a) > 0$; there exist uniquely determined vectors $x_i \in H_i$ such that $\|x_i\| = 1$, $(x_i|a_i) > 0$ and for each finite subset J of I :

$$x = (\bigotimes_{i\in J} x_i) \otimes x'_{(J)} \tag{A.5}$$

with

$$x'_{(J)} \in \overset{h}{\bigotimes}{}^{a}_{i\in I-J} H_i \, .$$

Denote by P the orthogonal projection onto x , by P_i the same onto x_i , and set

$$P_{(J)} = \bigotimes_{i\in J} P_i = \text{projection onto } \bigotimes_{i\in J} x_i$$

$$\overline{P}_{(J)} = P_{(J)} \otimes I \in \mathcal{L}(H) \quad ;$$

we shall prove that $\overline{P}_{(J)}$ converges strongly to P . Since it has norm 1 it suffices to prove that $\overline{P}_{(J)}(y)$ tends to $P(y)$ for a total set of elements y , for instance $y = \overline{Q}(x)$ with $\overline{Q} = Q \otimes I$, Q a bouded linear operator in $\overset{h}{\underset{i \in K}{\otimes}} H_i$, K a finite subset of I .

We then have for $J \supset K$, according to (A.5)

$$\overline{Q}(x) = (Q(\underset{i \in K}{\otimes} x_i)) \otimes (\underset{i \in J-K}{\otimes} x_i) \otimes x'_{(J)}$$

$$\overline{P}_{(J)}(\overline{Q}(x)) = (P_{(K)}(Q(\underset{i \in K}{\otimes} x_i))) \otimes (\underset{i \in J-K}{\otimes} x_i) \otimes x'_{(J)}$$

$$= \left[(Q(\underset{i \in K}{\otimes} x_i) | \underset{i \in K}{\otimes} x_i). \underset{i \in K}{\otimes} x_i\right] \otimes (\underset{i \in J-K}{\otimes} x_i) \otimes x'_{(J)}$$

$$= \left[(\overline{Q}(x)|x). \underset{i \in K}{\otimes} x_i\right] \otimes x'_{(K)}$$

$$= (\overline{Q}(x)|x).x = P(\overline{Q}(x)) .$$

We have thus proved that $\overline{P}_{(J)}$ converges strongly to P. In particular $\overline{P}_{(J)}(\otimes a_i)$ tends to $P(\otimes a_i)$, i.e.

$$(\underset{i \in J}{\otimes} (a_i|x_i).x_i) \otimes (\underset{i \in I-J}{\otimes} a_i) = \prod_{i \in J} (a_i|x_i)(\underset{i \in J}{\otimes} x_i) \otimes (\underset{i \in I-J}{\otimes} a_i$$

$$\longrightarrow (a|x).x \quad ; \tag{A.6}$$

taking the norms of both sides we get

$$\prod_{i \in J} (a_i|x_i) \longrightarrow (a|x) \quad ;$$

this implies (A.2) ; since (A.1) is trivially satisfied, $\otimes x_i$ exists and is equal to $\lim_{J}(\underset{i \in J}{\otimes} x_i) \otimes (\underset{i \in I-J}{\otimes} a_i)$ which is equal to x according to (A.6).

Proposition A.2. In a finite tensor product $H = \bigotimes_{i\in I}^{h} H_i$, the set of all decomposable vectors is closed.

Let $(x^{(n)})$ be a sequence of decomposable vectors converging to a vector x ; we can suppose $\|x\| = \|x^{(n)}\| = 1$; set $x^{(n)} = \otimes x_i^{(n)}$ with $\|x_i^{(n)}\| = 1$; there exists a partial sequence (n_p) such that for every i , $x_i^{(n_p)}$ has a weak limit x_i ; we shall prove that $x^{(n_p)}$ converges weakly to $\otimes x_i$, which will imply that $x = \otimes x_i$. Since $x^{(n_p)}$ is uniformly bounded it is sufficient to prove that $(x^{(n_p)} | a)$ tends to $(\otimes x_i | a)$ for a running over a total set, for instance $a = \otimes a_i$; then

$$(x^{(n_p)} | \otimes a_i) = \prod_i (x_i^{(n_p)} | a_i) \longrightarrow \prod_i (x_i | a_i) = (\otimes x_i | \otimes a_i).$$

Remark A.1. One can deduce from prop.A.1 and A.2 that the set of all decomposable vectors in an infinite tensor product is the closure of the set of all elementary decomposable vectors.

§ A.4. Tensor products of operators

Suppose we are given continuous linear operators T_i in H_i such that $T_i = I$ a.e.; there exists a unique operator in $H = \bigotimes_{i}^{h\,a} H_i$ denoted $\otimes T_i$, such that $(\otimes T_i)(\otimes x_i) = \otimes T_i x_i$ for every decomposable vector $\otimes x_i$. Consider now a family of operators $T_i^{(u)}$ such that for every u , $T_i^{(u)} = I$ a.e.; if for every i the family $(T_i^{(u)})$ is irreducible in H_i , then the family $(\otimes T_i^{(u)})$ is irreducible in H .

Appendix B. COHOMOLOGY OF GROUPS

§ B.1. General definitions

We refer to [27] for the general notion of cohomology.

Let G be a group and Γ a commutative group which we write additively ; suppose G acts by automorphisms in Γ, the transformed of an element γ of Γ by an element g of G being written $g.\gamma$. A 1-cocycle is a mapping u of G into Γ satisfying

$$u(gg') = u(g) - g.u(g') \quad ;$$

this implies $u(e) = 0$ and $u(g^{-1}) = - g^{-1}.u(g)$. The 1-cocycles constitute a group $Z^1(G,\Gamma)$. For every γ in Γ the mapping $g \longmapsto g.\gamma - \gamma$ is a 1-cocycle, called coboundary of γ and denoted by $\partial\gamma$. The kernel of ∂ is the set of all γ which are invariant under G ; its image is a subgroup $B^1(G,\Gamma)$ of $Z^1(G,\Gamma)$; the quotient $H^1(G,\Gamma) = Z^1(G,\Gamma)/B^1(G,\Gamma)$ is the first cohomology group. A 2-cocycle is a mapping v of $G \times G$ into Γ satisfying

$$g.v(g',g'') - v(gg',g'') + v(g,g'g'') - v(g,g') = 0 \; ;$$

for every mapping u of G into Γ the mapping

$$(g,g') \longmapsto u(gg') - u(g) - g.u(g')$$

is a 2-cocycle called coboundary of u and denoted by ∂u ; one defines analogously Z^2, B^2, H^2.

§ B.2. Study of Z^1 in the case of a unitary representation

(See [1],[15]). We suppose Γ is a Hilbert space H acted upon by a (continuous unitary) representation U of G and we consider only continuous 1-cocycles ; a 1-cocycle is continuous iff it is continuous at e. Every bounded 1-cocycle is a coboundary ; in particular if G is compact, every 1-cocycle is a coboundary.

We now look at the case where G is a separable commutative locally compact group, H is separable and U does not contain the trivial representation. By Stone's theorem we can desintegrate U into characters, i.e. write H as a Hilbert integral (see definition in [8] or [16]) over $\hat{G}$ with some measure μ :

$$H = \int_{\hat{G}}^{\oplus} H_{\chi} \, d\mu(\chi)$$

in such a manner that for every g in G, U_g becomes the multiplication operator by the function $\chi \longmapsto \langle \chi, g \rangle$; moreover the neutral element ε of G is μ-negligible. Denote by $\bar{H}$ the set of all (equivalence classes of) measurable vector fields $\omega = (\omega_{\chi})$ such that the integral $\int |\langle \chi, g \rangle - 1|^2 . \|\omega_{\chi}\|^2 . d\mu(\chi)$ is finite for every g in G and tends to 0 when g tends to e [for instance if $G = \underline{R}$ this amounts to saying that $\int \frac{\chi^2}{1+\chi^2} \|\omega_{\chi}\|^2 . d\mu(\chi)$ is finite] ; $\bar{H}$ is a linear space containing H . With each ω in $\bar{H}$ we can associate a 1-cocycle u , u(g) having coordinates $u(g)_{\chi}$ given by $u(g)_{\chi} = (\langle \chi, g \rangle - 1)\, \omega_{\chi}$; the mapping $\bar{\partial} : \omega \longmapsto u$ extends the mapping $\partial : H \longrightarrow Z^1$. It is proved in [15] that $\bar{\partial}$ is a bijection from $\bar{H}$ onto Z^1, and that for each ω in $\bar{H}$ and each neighbourhood V of ε in $\hat{G}$, we have $\int_{\hat{G}-V} \|\omega_{\chi}\|^2 . d\mu(\chi) < \infty$.

In particular if μ has no mass in some neighbourhood of ε (which is always the case if H is finite dimensional), $\bar{H}$ is equal to H and ∂ is bijective. More generally one can prove (see [15]) that $\bar{\partial}$ is bijective if G is locally compact and nilpotent, but of course $\bar{\partial}$ has to be defined in another way.

Appendix C. BOOLEAN ALGEBRAS

(See [19],[41])

§ C.1. General properties

Definition C.1. A Boolean algebra is a (partially) ordered set E satisfying the following conditions :

(i) any two elements a and b have a l.u.b. $a \vee b$ and a g.l.b. $a \wedge b$

(ii) E contains a smallest element $\mathbb{0}$ and a largest element $\mathbb{1}$

(iii) for every a in E there exists a unique element a', called the complement of a , such that $a \vee a' = \mathbb{1}$ and $a \wedge a' = \mathbb{0}$

(iv) distributivity law :

$$a \wedge (b \vee c) = (a \wedge b) \vee (a \wedge c)$$

$$a \vee (b \wedge c) = (a \vee b) \wedge (a \vee c)$$

(v) $(a \vee b)' = a' \wedge b'$, $(a \wedge b)' = a' \vee b'$.

Two elements a,b are said to be disjoint if $a \wedge b = 0$ or equivalently $a \leq b'$; E is complete (resp. σ-complete) if every subset (resp. countable subset) F of E has a l.u.b. and a g.l.b. ; if $F = \{a_i\}_{i \in I}$ they are denoted by $\bigvee_{i \in I} a_i$ and $\bigwedge_{i \in I} a_i$; if the a_i's are mutually disjoint, the family $(a_i)_{i \in I}$ is called a partition of $\vee a_i$. E is called σ-decomposable if every partition is countable ; every σ-complete σ-decomposable Boolean algebra is complete. An atom of E is a minimal non zero element a , which means that $b \leq a$ implies $b = a$ or $b = 0$; E is called atomic if every non zero element majorizes at least one atom ; and non atomic if it has no atoms. Given two Boolean algebras E and F a morphism of E into F is a mapping preserving the operations $\vee$, $\wedge$ and ' ; it is a σ -morphism if it preserves countable sup and inf.

Examples

1) The set $\mathcal{P}(X)$ of all subsets of a set X is a Boolean algebra for the ordinary operations ; the Boolean algebras which arise in this way are exactly the complete atomic ones. Every σ -complete sub-Boolean algebra of $\mathcal{P}(X)$ is called a σ -algebra of subsets of X.

2) The set of all open-closed subsets of a topological space X is a Boolean algebra for the ordinary operations ; conversely (Stone's representation theorem) every Boolean algebra is of this type with a totally disconnected compact X (cf.[19],[16])

3) Let (X, μ) be a measure space with finite positive μ ; the set of all equivalence classes of measurable subsets is a complete σ -decomposable Boolean algebra ; it is non atomic iff μ is non atomic, i.e. iff every point in X is μ -negligible.

4) Loomis theorem : every σ -complete Boolean algebra is the image of some σ -algebra of subsets of a set under some σ -morphism.

Lemma C.1. Let E be a complete Boolean algebra, F a subset satisfying the following conditions :

(i) if elements a_i belong to F and are mutually disjoint, then $\vee a_i \in F$

(ii) $a \in F$ and $b \leq a$ imply $b \in F$.

Then the l.u.b. of F belongs to F.

The proof can be found in [41], § 21.

Lemma C.2. Let E be a nonatomic complete σ-decomposable Boolean algebra and φ a non decreasing positive real valued function on E satisfying $\varphi(\vee a_i) \leq \Sigma\varphi(a_i) < \infty$ for every partition (a_i). Then for every $\varepsilon > 0$ there exists a finite partition of $\mathbb{1}$: $\mathbb{1} = \vee b_i$ such that $\varphi(b_i) \leq \varepsilon \quad \forall i$.

Set $G = \{a \mid \varphi(a) \leq \varepsilon\}$ and $c = \sup_{a \in G} a$. If $c \neq \mathbb{1}$, c' has an infinite partition $c' = \vee d_i$ and we have $d_i \notin G$, $\varphi(d_i) > \varepsilon \quad \forall i$, which is impossible. Thus $c = \mathbb{1}$. Let $(b_j)_{j \in J}$ a maximal disjoint family in G ; if $\vee b_j \neq \mathbb{1}$ there exists an element a in G which is not majorized by $\vee b_j$; then $a \wedge (\vee b_j)'$ belongs to G (since φ is non decreasing) and is disjoint from each b_j, which contradicts the maximality of (b_j). Thus $\vee b_j = \mathbb{1}$. Since $\Sigma \varphi(b_j)$ is finite there exists a finite subset $J' \subset J$ such that $\sum_{j \in J - J'} \varphi(b_j) \leq \varepsilon$; then

$$\varphi\Big(\bigvee_{j \in J - J'} b_j\Big) \leq \sum_{j \in J - J'} \varphi(b_j) \leq \varepsilon$$

and we can take for (b_i) the family $(b_j)_{j \in J'}$ plus the element $\bigvee_{j \in J - J'} b_j$.

§ C.2. σ-additive and σ-multiplicative functions

Definition C.2. A complex function f on a σ-complete Boolean algebra E will be called σ-*additive* (resp. σ-multiplicative) if for every countable partition $a = \vee a_i$ we have $\sum f(a_i) = f(a)$ (resp. $\sum |f(a_i) - 1| < \infty$ and $\prod f(a_i) = f(a)$).

Proposition C.1. Let E be a non atomic complete σ-decomposable Boolean algebra and f a σ-multiplicative complex function on E. There exists a unique σ-additive function g such that we have $f(a) = e^{g(a)} \quad \forall a \in E$.

Proof.

a) We first prove that $f(a) \neq 0 \ \forall a$. Let $F = \{a \mid f(a) \neq 0\}$; then $b \leq a \in F$ implies $b \in F$; moreover F is stable for the sup of disjoint families; by lemma C.1 the l.u.b. c of F belongs to F. If $c \neq \mathbb{1}$, c' admits an infinite partition $c' = \vee d_i$ and we have $d_i \notin F$, $f(d_i) = 0 \ \forall i$ which is impossible. Thus $c = \mathbb{1}$, $f(\mathbb{1}) \neq 0$ and $f(a) \neq 0 \ \forall a$.

b) We shall prove the existence of g in the case where $|f(a)| \leq 1$ for every a. Set $\varphi(a) = \sup_{b \leq a} |f(b) - 1|$; φ is non decreasing. Moreover for every partition $a = \bigvee_{i=1}^{\infty} a_i$ we have

$$\varphi(a) \leq \sum \varphi(a_i) \tag{C.1}$$

$$\sum \varphi(a_i) < \infty \tag{C.2}$$

In fact take an element $b \leq a$ and set $b_i = b \wedge a_i$; then

$$f(b) = \prod f(b_i)$$

$$|f(b) - 1| = |\prod f(b_i) - 1| \leq \sum |f(b_i) - 1|$$

[in fact for every countable family of complex numbers z_i with $|z_i| \leq 1$ we have $|\prod z_i - 1| \leq \sum |z_i - 1|$]; then $|f(b) - 1| \leq \sum \varphi(a_i)$ and since this holds for every $b \leq a$, we get (C.1).
To prove (C.2) we can choose for every i an element $b_i \leq a_i$ such that

$$\varphi(a_i) \leq |f(b_i) - 1| + 2^{-i}$$

then

$$\sum \varphi(a_i) \leq \sum |f(b_i) - 1| + 1$$

and the right hand side is finite since the b_i's are mutually disjoint ; this proves (C.2).
We can now apply lemma C.2 : there exists a finite partition $\mathbb{1} = \vee b_i$ such that $\varphi(b_i) \leq \frac{1}{2}$ $\forall i$; for every b majorized by some b_i we have $|f(b) - 1| \leq \frac{1}{2}$ and we can consider the principal determination $\log f(b)$; now if a is an arbitrary element of E we define $g(a)$ as $\sum \log f(a \wedge b_i)$ and g has the required properties.

c) In this section we prove the existence of g in the general case. By part a) we can consider the real function $\psi(a) = \log|f(a)|$ which satisfies for every partition $a = \vee a_i$: $\psi(a) = \sum \psi(a_i)$.
There exists an element c in E such that

$$\psi(a) \leq 0 \quad \forall a \leq c$$

$$\psi(a) \geq 0 \quad \forall a \leq c'$$

[this can be proved exactly as the corresponding result in classical measure theory]; E splits into two parts $E_1 = \{a \mid a \leq c\}$,

$E_2 = \{a \mid a \leq c'\}$; on E_1 we have $|f(a)| \leq 1$ and we can apply part b) of the proof ; on E_2 we have $|f(a)| \geq 1$ and we can apply b) to the function $1/f$.

d) Unicity : suppose g and g' both satisfy the conditions of the proposition ; then $h = (g-g')/2\pi i$ has entire values and satisfies $\sum h(a_i) = h(a)$ for every partition ; by the proof of c) we can suppose $h \geq 0$; lemma C.2 implies $h(b_i) = 0$, whence $h(a) = 0 \quad \forall a$.

§ C.3. Multiplicative measures

Definition C.3. A σ-multiplicative measure on a Borel space T with Borel structure $\mathcal{B}$ is a mapping $\tau : \mathcal{B} \longrightarrow \underline{C}^*$ such that for every countable partition $S = \bigcup_{i \in I} S_i$ with S and $S_i \in \mathcal{B}$, we have $\sum |\tau(S_i) - 1| < \infty$ and $\prod \tau(S_i) = \tau(S)$.

Proposition C.2. If τ is a σ-multiplicative measure on T such that $\tau(\{t\}) = 1 \quad \forall t \in T$, there exists a unique σ-additive measure $\mu : \mathcal{B} \longrightarrow \underline{C}$ such that $\tau(S) = e^{\mu(S)} \quad \forall S \in \mathcal{B}$.

Let us denote by $\mathcal{B}^0$ the set of all S in $\mathcal{B}$ such that

$$\tau(S') = 1 \quad \forall S' \in \mathcal{B} , S' \subset S ;$$

identifying two subsets in $\mathcal{B}$ which differ by a subset of $\mathcal{B}^0$ we get a Boolean algebra E which is non atomic (the proof is similar to a classical one in ordinary measure theory), σ-complete and σ-decomposable, hence complete ; τ defines a σ-multiplicative function on E and we can apply prop.C.1.

Appendix D. RESTRICTED PRODUCTS OF SETS AND MEASURES

(See [14])

§ D.1. Restricted products of sets and Borel spaces

Definition D.1. Given a family $(X_i)_{i \in I}$ of sets and for each i a subset Y_i of X_i we call restricted product of the family (X_i) with respect to the family (Y_i) the set of all families (x_i) in the cartesian product $\prod X_i$ such that $x_i \in Y_i$ for almost all i. We denote it by $\prod_{i \in I}^{(Y_i)} X_i$. If Y_i is reduced to a single point a_i we write $\prod^{(a_i)} X_i$ instead of $\prod^{(\{a_i\})} X_i$.

For every finite subset J of I we denote by $X_{(J)}$ the set of all families (x_i) such that $x \notin J \Longrightarrow x_i \in Y_i$, i.e.

$$X_{(J)} = \prod_{i \in J} X_i \times \prod_{i \in I-J} Y_i ;$$

these sets form a non decreasing family and $\prod^{(Y_i)} X_i$ is their union.

Restricted products of sets have associativity and commutativity properties analogous to those of infinite tensor products of Hilbert spaces (see App.A.2).

Example D.1. Suppose that each X_i is a CMU (commutative monoid with unit) with unit element 0_i ; $\prod^{(0_i)} X_i$ is simply written $\prod' X_i$ and called restricted product of the X_i's ; as is known $\prod' X_i$ is the sum of the X_i's in the category of all CMU.

Definition D.2. Suppose now that every X_i is a Borel space and Y_i a Borel subset ; we endow each $X_{(J)}$ with the product Borel structure ; then for $J \subset K$, $X_{(J)}$ is a Borel subspace of $X_{(K)}$; we define a Borel structure on the restricted product X by saying that a subset U of X is Borel iff for every J, $U \cap X_{(J)}$ is Borel in $X_{(J)}$. Then each $X_{(J)}$ appears as a Borel subspace of X ; the Borel space X is called Borel restricted product of the X_i's.

If I is countable and each X_i is standard there is an alternative definition of the Borel structure in X. In fact we have

Proposition D.1. With the above assumptions denote by $\mathcal{A}$ (resp. $\mathcal{A}_i$) the set of all Borel subsets in X (resp. X_i) and by p_i the projection $X \longrightarrow X_i$. Then $\mathcal{A}$ is generated by the σ-algebras $p_i^{-1}(\mathcal{A}_i)$.

Clearly the σ-algebra $\mathcal{A}'$ generated by the $p_i^{-1}(\mathcal{A}_i)$ is contained in $\mathcal{A}$; since the Borel structure defined by $\mathcal{A}$ is standard, it is sufficient to prove that $\mathcal{A}'$ is generated by a sequence of subsets which, moreover, separate the points of X (see [9] App.B or [32]). Let $(A_{i,n})$ be a countable family generating $\mathcal{A}_i$; if we have two distinct points in X, say $x \neq x'$, there exists i_o such that $x_{i_o} \neq x'_{i_o}$; there exists n such that $x_{i_o} \in A_{i_o,n}$ and $x'_{i_o} \notin A_{i_o,n}$; then x belongs to $p_{i_o}^{-1}(A_{i_o,n})$ while x' does not belong to it. We thus see that the subsets $p_i^{-1}(A_{i,n})$ separate the points ; moreover they clearly generate $\mathcal{A}'$.

Example D.2. A Borel commutative monoid with unit is a CMU with a Borel structure such that the mapping $(x,y) \longmapsto x+y$ of $X \times X$ into X is Borel. If $(X_i)_{i \in I}$ is a family of Borel CMU's , the restricted product $\prod' X_i$ is the categorical sum of the X_i's in the category of Borel CMU.

§ D.2. Measure spaces

We call measure space a pair (X, μ) where X is a Borel space and μ a finite positive Borel measure on X. Two measure spaces (X, μ) and (X', μ') are said to be isomorphic if there exist a μ-negligible subset Y in X, a μ'-negligible subset Y' in X' and a μ-measurable bijective mapping u of $X-Y$ onto $X'-Y'$ such that $u(\mu)$ is equivalent to μ' ; such an isomorphism will be called measure preserving if $u(\mu) = \mu'$.

With evey measure space one can associate a Boolean algebra by taking the σ-algebra of all measurable subsets and identifying two subsets which differ by a negligible subset . One can also associate the Hilbert space $L^2(X, \mu)$ and the von Neumann algebra $\mathcal{A}(X, \mu)$ in $L^2(X, \mu)$, the elements of which are the multiplication operators by functions in $L^\infty(X, \mu)$ (these operators are also called diagonalisable operators) ; then every isomorphism $u : (X, \mu) \longrightarrow (X', \mu')$ gives rise to an isomorphism $A : L^2(X, \mu) \longrightarrow L^2(X', \mu')$ carrying $\mathcal{A}(X, \mu)$ onto $\mathcal{A}(X', \mu')$, namely

$$(Af)(x') = \left[\frac{d\mu'(x')}{d\mu(x)}\right]^{\frac{1}{2}} . f(u^{-1}(x')) \quad \forall \; f \in L^2(X, \mu).$$

Conversely every isomorphism $L^2(X,\mu) \longrightarrow L^2(X',\mu')$ carrying $\mathcal{A}(X,\mu)$ onto $\mathcal{A}(X',\mu')$ arises in this manner from an isomorphism $(X,\mu) \longrightarrow (X',\mu')$ (see [8]).

The von Neumann algebra $\mathcal{A}(X,\mu)$ associated with (X,μ) has a cyclic veetor, namely the function 1 ; conversely given a commutative von Neumann algebra in a Hilbert space H with a cyclic vector ξ there exist a measure space (X,μ) and an isomorphism $H \longrightarrow L^2(X,\mu)$ carrying ξ into the function 1 and $\mathcal{A}$ onto the diagonalisable operators. In particular if G is a commutative group and φ a positive definite function on G one can realize H_φ as a space $L^2(X,\mu)$, ξ_φ as the function 1 and $U_\varphi(g)$ as the multiplication operator by some function of modulus 1 (see definition of H_φ, ξ_φ, U_φ in §3.1) ; moreover the measure space (X,μ) is unique up to isomorphism.

§ D.3. Restricted products of measure spaces

Definition D.3. We keep the notations of § D.1 and suppose that on each X_i we have a (positive Borel) measure μ_i satisfying $\mu_i(Y_i)=1$ (we emphasize that in general $\mu_i(X_i) > 1$) ; we set $\nu_i = \mu_i | Y_i$; for each finite subset J in I we can form the measure $\mu_{(J)} = (\bigotimes_{i\in J} \mu_i) \otimes (\bigotimes_{i\in I-J} \nu_i)$ on the space $X_{(J)}$; for $J \subset K$ we have

$$\mu_{(K)} | X_{(J)} = \mu_{(J)} .$$

Hence there exists a unique measure μ on X such that we have

$$\mu | X_{(J)} = \mu_{(J)} \quad \forall J ;$$

μ will be called the <u>restricted product</u> of the μ_i's and denoted by $\bigotimes_{i\in I}^{(Y_i)} \mu_i$; the measure space (X,μ) is denoted $\prod_{i\in I}^{(Y_i)}(X_i,\mu_i)$.

If $Y_i = X_i$ this is nothing but the ordinary product. We have

$$\|\mu\| = \prod_{i\in I} \|\mu\|_i \leq +\infty .$$

<u>Relation with infinite tensor products of Hilbert spaces</u>

Let a_i be the characteristic function of Y_i ; it is a unit vector in $L^2(X_i,\mu_i)$; there exists a unique isomorphism T of $\overset{h}{\otimes}{}^{(a_i)} L^2(X_i,\mu_i)$ onto $L^2(X,\mu)$ with the following property : for every family (f_i) where $f_i \in L^2(X_i,\mu_i)$ and $f_i = a_i$ a.e., $T(\otimes f_i)$ is the function on X : $(x_i) \longmapsto \prod_{i\in I} f_i(x_i)$.(see [14]).

Appendix E. GAUSSIAN AND POISSONIAN MEASURES AND PROCESSES

§ E.1. Gaussian measures on $\underline{R}^n$

A probability measure μ on $\underline{R}^n$ is called Gaussian if it is absolutely continuous with respect to Lebesgue measure with a density of the form $e^{-Q(x)}$ where Q is a non degenerate positive quadratic form ; more precisely a Gaussian measure can be written

$$d\mu(x) = (2\pi)^{-n/2} (\det A)^{\frac{1}{2}} e^{-(Ax|x)/2} dx$$

where A is an invertible positive definite linear operator in $\underline{R}^n$ and $(x|y)$ the usual scalar product in $\underline{R}^n$. Its characteristic function is

$$\varphi(t) = \int e^{i(t|x)} d\mu(x) = e^{-(A^{-1}t|t)/2} .$$

In particular for $A = I$ we get the so called reduced Gaussian measure.

We recall some useful classical formulas. For every $a > 0$ and every complex t we have

$$\int_{-\infty}^{+\infty} e^{itx-ax^2/2} dx = (2\pi/a)^{\frac{1}{2}} e^{-t^2/2a} \quad ;$$

for every integer $n \geq 0$ we have

$$\int_{-\infty}^{+\infty} x^{2n} e^{-x^2/2} dx = (2\pi)^{\frac{1}{2}} (2n) /2^n n!$$

$$\int_{-\infty}^{+\infty} x^{2n-1} e^{-x^2/2} dx = 0 .$$

The Hermite polynomials h_n are defined by

$$e^{ux-u^2/2} = \sum_{n=o}^{\infty} u^n h_n(x)/n!$$

$$h_n(x) = e^{x^2/2} (-1)^n \frac{d^n}{dx^n} e^{-x^2/2} ;$$

the functions $(n\,)^{-\frac{1}{2}} h_n$ form an orthonormal basis of $L^2(\underline{R}, \nu)$ where ν is the reduced Gaussian measure on $\underline{R}$.

§ E.2. Poissonian measures on R

A probability measure μ on $\underline{R}$ is called Poissonian (or mixed Poissonian) if its characteristic function has the form

$$\varphi(t) = \exp\left[\int_{\underline{R}} (e^{ikt} - 1)\, d\nu(k) + iat\right] \tag{E.1}$$

where ν is a finite positive measure on $\underline{R}$ and a a real number. If ν is concentrated on a single point, say $\nu = \lambda\delta_k$, μ is said <u>pure Poissonian</u> ; then

$$\varphi(t) = \exp\left[\lambda(e^{ikt} - 1) + iat\right]$$

and μ is defined by the mass $e^{-\lambda}\lambda^n/n!$ at each point $kn+a$, $n = o,1,2,\ldots$ One has also to consider Poissonian measures of a more general type, namely

$$\varphi(t) = \exp\left[\int_{\underline{R}} (e^{ikt} - 1 - ikt/(1+k^2))\, d\nu(k) + iat\right] \tag{E.2}$$

where ν is a positive measure on $\underline{R} - \{0\}$ such that we have $\int k^2(1+k^2)^{-1}\, d\nu(k) < \infty$; if ν is bounded we have again (E.1).

§ E.3. Stochastic processes

A <u>stochastic process</u> on a set E is a mapping associating with each element a of E an equivalence class X_a of real measurable functions (in other words X_a is a real random variable) on a probability space (Ω, P). Let μ_a be the image of P under X_a, φ_a the characteristic function of μ_a :

$$\varphi_a(t) = \int e^{itx} \, d\mu_a(x) = P(e^{itX_a}) .$$

If X_a is square integrable for every a , we can consider the kernel on E : $(a,b) \longrightarrow P(X_a X_b)$; it is called the covariance of the process X.

Now suppose that E is a group and that X is a morphism, i.e. $X_{ab} = X_a + X_b$; then we can define a unitary representation U_t of E in $L^2(\Omega, P)$ for every real t , $U_t(a)$ being the multiplication operator by e^{itX_a} ; we have

$$\varphi_a(t) = P(e^{itX_a}) = (U_t(a).1 \mid 1)$$

which proves that the function $a \longmapsto \varphi_a(t)$ is positive definite. In particular suppose that E is a real vector space and that X is linear ; then $\varphi_a(t) = \varphi_{ta}(1)$ and it is sufficient to consider $\varphi_a(1)$; we write it $\varphi(a)$ and we see that φ is a positive definite function on E ; it is called the <u>characteristic functional</u> of X. If X_a is square integrable for every a , the covariance is a bilinear functional on E.

A very important particular case of linear process is that where E is the set of all real Borel functions on a standard Borel space T which take only a finite number of values ; then a linear process X on E is called <u>decomposable</u> if (writing X_A instead of X_{χ_A} for A Borel subset of T) for every sequence of disjoint Borel subsets $A_1, A_2, \ldots$, the random variables $X_{A_1}, X_{A_2}, \ldots$ are independant and the series $\sum X_{A_n}$ converges almost everywhere to $X_{\cup A_n}$. We could deduce from chapter 6 some information about the characteristic functional φ of a decomposable linear process ; but one can prove (see e.g. [11]) that if $X_{\{t\}} = 0$ for every t in T , φ has the form

$$\varphi(f) = \exp\Big[i\int f(t).d\alpha(t) - \int f(t)^2.d\beta(t) + \\ + \int (e^{i\lambda f(t)} - 1 - i\lambda f(t)/(1+\lambda^2)).d\gamma(t,\lambda)\Big]$$

where α is a finite real measure on T , β a finite positive one, γ a positive measure on $T \times \underline{R}^*$ such that

$$\int_{T\times \underline{R}^*} \lambda^2/(1+\lambda^2).d\gamma(t,\lambda) < \infty .$$

X and φ are called <u>Gaussian</u> if γ is null and <u>Poissonian</u> if β is null.

Appendix F. CANONICAL COMMUTATION RELATIONS

(See [13])

Definition F.1. Let E be a separated complex prehilbert space with scalar product $(x|y)$; a representation of the canonical commutation relations over E (or in short representation of E) in a complex Hilbert space H is a mapping U of E into $\mathcal{U}(H)$ (set of all unitary operators in H) satisfying the following conditions :

(i) $U(x + y) = \exp(-i \operatorname{Im}(x|y))\, U(x)\, U(y)$

(ii) the restriction of U to an arbitrary finite dimensional linear subspace of E is strongly continuous.

Condition (i) means that U is a projective representation of the additive group E with multiplier $(x,y) \longmapsto \exp(-i \operatorname{Im}(x|y))$. in the sense of [26].

If ξ is a unit vector in H the function φ on E defined by $\varphi(x) = (U(x)\xi|\xi)$ satisfies the following conditions :

(iii) $\varphi(0) = 1$

(iv) for any $x_1, \dots x_n$ in E the matrix with coefficients $\exp(-i \operatorname{Im}(x_p|x_q)).\varphi(x_p - x_q)$, $p,q = 1,\dots n$, is positive definite

(v) φ is continuous on every finite dimensional linear subspace of E .

Conversely there is a construction similar to the Gelfand-Segal construction (see § 3.1) : with every φ satisfying (iii),(iv),(v) one can associate canonically a Hilbert space H_φ , a unit vector ξ_φ and a representation U_φ such that $\varphi(x) = (U_\varphi(x).\xi_\varphi \mid \xi_\varphi)$; the scalar product in $\underline{C}^{(E)}$ is defined by

$$(f|f') = \sum_{x,x'} \exp(i\ \mathrm{Im}(x|x')).\varphi(x-x').f(x).\overline{f'(x')}$$

and the representation of E in $\underline{C}^{(E)}$ by

$$(U^{o}(x).f)(y) = \exp(i\ \mathrm{Im}(x|y)).f(y-x).$$

For instance the function $x \longmapsto \exp(-\|x\|^2/2)$ satisfies (iii), (iv),(v) ; the associated representation, called the <u>Fock representation</u>, is described in § 2.2 ; it acts on $S\bar{E}$ where $\bar{E}$ is the completion of E ; $S\bar{E}$ is also called the <u>Fock space</u> of E. It follows from prop.2.3 and 2.5 that the Fock representation of a direct sum of spaces E_i is the tensor product of the Fock representations of the spaces E_i ; physicists often use this result to give a concrete description of the Fock representation of a general E : one takes an orthonormal basis of E and consider E as a direct sum of one-dimensional spaces ; of course this construction is not intrinsic. We show in remark 8.2 that when E is a space of functions one can sometimes consider intrinsically its Fock representation as a continuous tensor product of Fock representations of one-dimensional spaces.

Appendix G. MEASURES ON TOPOLOGICAL VECTOR SPACES

§ G.1. General properties

Definition G.1. A real topological vector space E is said to be a separable $(\mathcal{L}\mathcal{F})$-space if it contains a non decreasing sequence of linear subspaces E_n , $n = 1,2,...$ such that

(i) E is the union of the E_n's ;

(ii) for the induced topology each E_n is a Fréchet space (i.e. locally convex, metrisable and complete) and moreover separable ;

(iii) if a subset U of E is convex and if $E_n \cap U$ is a neighbourhood of 0 in E_n for every n, then U is a neighbourhood of 0 in E .

Then E is separable. We denote by $\mathcal{F}(E)$ the set of all finite dimensional linear subspaces of E . It can be proved that a positive definite function on E is continuous iff it is continuous on each E_n (see [5],§6).

The topological dual E' of E will always be endowed with the weak topology ; the underlying Borel space is standard (see [5], § 5) ; let $\mathcal{A}$ be the set of all Borel subsets. For every x in E we define functions $\tilde{x}$ and $\hat{x}$ on E' by

$$\tilde{x}(\chi) = \langle \chi , x \rangle \quad , \quad \hat{x}(\chi) = e^{i\langle \chi , x \rangle} \qquad \forall \chi \in E'.$$

For every F in $\mathcal{F}(E)$ we denote by Λ_F the canonical mapping of E' onto F' and by $\mathcal{A}_F$ the sub-σ-algebra of $\mathcal{A}$ which is the inverse image under Λ_F of the canonical σ-algebra on F' (a finite dimensional vector space has a canonical topology and therefore a canonical Borel structure) ; the $\mathcal{A}_F$'s form a filtering family which generates $\mathcal{A}$ (see [5], § 5) ; it follows that $\mathcal{A}$ is the σ-algebra defined by the functions $\tilde{x}$, $x \in E$ (i.e. the smallest σ-algebra for which all functions $\tilde{x}$ are measurable). The subsets belonging to the union of the $\mathcal{A}_F$'s are also called "cylindrical subsets".

We call <u>measure on</u> E' a finite positive Borel measure ; by [5] § 3 these measures are also the finite positive Radon measures. For every measure μ on E' the function on E :

$$x \longmapsto \mathcal{F}\mu(x) = \mu(\hat{x}) = \int e^{i<\chi,x>} . d\mu(\chi)$$

is positive definite and moreover continuous : in fact Lebesgue convergence theorem proves that $\mathcal{F}\mu$ is continuous on every E_n , and as we know already this implies that it is continuous on E. It is called the <u>Fourier transform</u> of μ . The mapping $\mathcal{F}$ is injective (see [4], § 6, prop.3) but not surjective in general ; however if E is nuclear every continuous p.d.f. on E is the Fourier transform of a measure (Bochner-Minlos theorem). When E is a separable Hilbert space we have the following characterization : a p.d.f. on E is a Fourier transform iff it is continuous for the topology defined by the semi-norms of the form $Q^{\frac{1}{2}}$ where Q is a positive quadratic form satisfying the following condition : there exists a constant k such that $\sum_{i=1}^{n} Q(e_i) \leq k$ for every orthonormal family $e_1, \ldots e_n$. In particular if A is a continuous linear operator in E the p.d.f. $\exp(-\|Ax\|^2)$ is a Fourier transform iff A is of Hilbert-Schmidt type ; this shows that $\exp(-\|x\|^2)$ is <u>never a Fourier transform</u> if E is infinite dimensional.

<u>Lemma</u> G.1. Let μ be a measure on E' ; for every $p \in [1,+\infty]$ the functions $\hat{x}$, $x \in E$, are total in $L^p(E',\mu)$ (for $p = \infty$ we take the weak topology of dual of L^1).

This result is classical for a finite dimensional E ; in the general case, for every F in $\mathcal{F}(E)$ the functions $\hat{x}$, $x \in F$, are total in $L^p(E', \mathcal{Q}_F, \mu|\mathcal{Q}_F)$ since this space can be identified with $L^p(F', \Lambda_F(\mu))$; then the lemma follows from the martingale theorem (see App.H.2).

<u>Relation with unitary representations of E</u>

Let μ be a normalized measure on E' and φ its Fourier transform ; then the representation of E associated with φ (see § 3.1) can be realized as follows :

$$H_\varphi = L^2(E',\mu)$$

$$\xi_\varphi = \text{function } 1$$

$$U_\varphi(x) = \text{multiplication operator by } \hat{x} \text{ ;}$$

in fact this defines a unitaty representation of E in $L^2(E',\mu)$, the function 1 is cyclic by the preceding lemma, and we have

$$(\hat{x}.1|1) = \mu(\hat{x}) = \varphi(x) .$$

§ G.2. Relation with linear stochastic processes

Let μ be a measure on E' ; the mapping $x \longmapsto X_x = \tilde{x}$ is a linear stochastic process (see definition in App.E) with characteristic functional $\varphi = \mathcal{F}\mu$; moreover we know that the functions X_x , $x \in E$, define the Borel structure of E'.

Conversely suppose that we are given a probability space $(\Omega, \mathcal{A}, P)$ and a linear stochastic process X on E with continuous characteristic functional φ ; let F be a dense subspace of E which is nuclear for a stronger topology [for instance F is a subspace algebraically generated by a countable dense subset and its topology is the finest locally convex topology (see § 7.1)], μ a normalized measure on F' with Fourier transform $\varphi | F$; let $\mathcal{B}$ be the sub-σ-algebra of $\mathcal{A}$ defined by the functions X_x , $x \in E$; then the functions e^{iX_x} , $x \in E$, are total in $L^2(\Omega, \mathcal{B}, P')$ where $P' = P|\mathcal{B}$ [to see this it is sufficient to show that the von Neumann algebra $\mathcal{U}$ generated by all multiplication operators by functions e^{iX_x} is equal to the von Neumann algebra of all multiplication operators by functions in $L^\infty(\Omega, \mathcal{B}, P')$; denote by $\mathcal{V}$ the last algebra and suppose that $\mathcal{U} \neq \mathcal{V}$; it is known that the elements of $\mathcal{U}$ are the multiplication operators by functions in $L^\infty(\Omega, \mathcal{C}, P|\mathcal{C})$ where $\mathcal{C}$ is some sub-σ-algebra of $\mathcal{B}$; but since the functions X_x define the σ-algebra $\mathcal{B}$, we have $\mathcal{B} = \mathcal{C}$ whence $\mathcal{U} = \mathcal{V}$].
Therefore the function 1 is cyclic for the representation V of F in $L^2(\Omega, \mathcal{B}, P')$ defined by

$$V(x).f \;=\; e^{iX_x} f \qquad \forall\, f \in L^2(\Omega\,, \mathcal{B}, P')\;;$$

on the other hand we have a representation U of F in $L^2(F', \mu)$:
$U(x)$ = multiplication operator by $\hat{x}$ and

$$(V(x).1|1) \;=\; P(e^{iX_x}) \;=\; \varphi(x) \;=\; \mu(\,\hat{x}\,) \;=\; (U(x).1|1)\;;$$

by App.D.2 there exists a measure preserving isomorphism $(\Omega\,, \mathcal{B}, P')$ $\longrightarrow (F', \mu)$ carrying e^{iX_x} into $\hat{x}$ $\forall\, x \in F$; since X_x and $\tilde{x}$ are linear in x , this isomorphism also carries X_x into $\tilde{x}$. This proves that <u>the linear stochastic process</u> X <u>restricted to</u> F <u>can be realized as</u> $x \longmapsto \tilde{x}$ <u>with a measure on</u> F'.

Appendix H. CONDITIONAL EXPECTATIONS AND MARTINGALE THEOREM

§ H.1. Conditional expectations

(See for instance [28]).

We consider a set X, a σ-algebra $\mathcal{A}$ of subsets of X, a probability measure μ on $\mathcal{A}$, a sub-σ-algebra $\mathcal{B}$ of $\mathcal{A}$ and the restriction $\mu' = \mu|\mathcal{B}$ of μ to $\mathcal{B}$. For every $p \in [1,+\infty]$ we can consider $L^p(X,\mathcal{B},\mu')$ as a subspace of $L^p(X,\mathcal{A},\mu)$. The <u>conditional expectation</u> is a mapping of $L^1(X,\mathcal{A},\mu)$ into $L^1(X,\mathcal{B},\mu')$ which is denoted by E or $E^{\mathcal{B}}$ and satisfies the following conditions :

(i) E is linear, continuous and with norm ≤ 1 ;

(ii) if $f \in L^1_+(X,\mathcal{A},\mu)$, Ef is the Radon-Nikodym derivative of $(f\mu)|\mathcal{B}$ with respect to $\mu|\mathcal{B}$;

(iii) $\mu(Ef.g) = \mu(fg)$ for every $g \in L^\infty(X,\mathcal{B},\mu')$ (this is a characterization of Ef) ;

(iv) if f belongs to $L^1(X,\mathcal{B},\mu')$ then $Ef = f$;

(v) if f belongs to $L^p(X,\mathcal{A},\mu)$ where $p \in [1,+\infty]$ then Ef belongs to $L^p(X,\mathcal{B},\mu')$ and $\|Ef\|_p \leq \|f\|_p$;

(vi) the restriction of E to $L^2(X,\mathcal{A},\mu)$ is the orthogonal projection of $L^2(X,\mathcal{A},\mu)$ onto $L^2(X,\mathcal{B},\mu')$;

(vii) if f belongs to $L^1_+(X,\mathcal{A},\mu)$ then $E(f^{\frac{1}{2}}) \leq (Ef)^{\frac{1}{2}}$.

<u>Example</u>. Let X' be a set, $\mathcal{A}'$ a σ-algebra on X', T a mapping of X into X' such that $T^{-1}(\mathcal{A}') \subset \mathcal{A}$, $\mathcal{B} = T^{-1}(\mathcal{A}')$, $T(\mu)$ the

image of μ under T , $f \in L^1_+ (X, \mathcal{A}, \mu)$, g the density of $T(f\mu)$ with respect to $T(\mu)$. Then $g \circ T = E^{\mathcal{B}} f$.

§ H.2. The martingale theorem

Let X be a set, $\mathcal{A}$ a σ-algebra on X, μ a probability measure on $\mathcal{A}$, I a filtering ordered set, $(\mathcal{A}_i)_{i \in I}$ a non decreasing family of sub-σ-algebras generating $\mathcal{A}$. For every f in $L^p(X, \mathcal{A}, \mu)$ where $p \in [1, +\infty]$, $E^{\mathcal{A}_i} f$ converges to f in the L^p sense (for $p = \infty$ we take the weak topology of dual of L^1). If moreover $I = \underline{N}$ with its usual ordering, for every f in $L^1(X, \mathcal{A}, \mu)$, $E^{\mathcal{A}_i} f$ converges almost everywhere to f.

§ H.3. Hellinger integrals and equivalence of measures

(See e.g. [30], lemma 8.2)

If μ and ν are two probability measures on a Borel space $(X, \mathcal{A})$, one defines their <u>Hellinger integral</u> $\int \sqrt{d\mu . d\nu}$ by

$$\int \sqrt{d\mu . d\nu} = \int f^{\frac{1}{2}} g^{\frac{1}{2}} d\pi$$

where π is an arbitrary measure dominating μ and ν (which means that μ and ν are absolutely continuous with respect to π) and f and g the Radon-Nikodym derivatives of μ and ν with respect to π . It is null iff μ and ν are disjoint. If μ' and ν' are the restrictions of μ and ν to some sub-σ-algebra of $\mathcal{A}$ we have

$$\int \sqrt{d\mu' . d\nu'} \geqslant \int \sqrt{d\mu . d\nu} .$$

Let us now consider a filtering family of sub-σ-algebras $(\mathcal{A}_i)_{i \in I}$

generating $\mathcal{A}$; let μ_i , ν_i be the restrictions of μ , ν to $\mathcal{A}_i$; then one can prove the following : the family of positive numbers $\int\sqrt{d\mu_i . d\nu_i}$ is non increasing, and its limit is zero iff μ and ν are disjoint. In particular suppose $(X, \mathcal{A})$ is the projective limit of a family $(X_i, \mathcal{A}_i)$ with mappings $\Lambda_i : X \longrightarrow X_i$; set $\mathcal{A}_i = \Lambda_i^{-1}(\mathcal{B}_i)$; let μ and ν be two probability measures on $(X, \mathcal{A})$, μ_i' and ν_i' their projections onto X_i ; then the family of numbers $\int\sqrt{d\mu_i' . d\nu_i'}$ is non increasing and its limit is zero iff μ and ν are disjoint.

More particularly suppose that $(X, \mathcal{A})$ is a product

$$(X, \mathcal{A}) = \prod_{n=1}^{\infty} (X^{(n)}, \mathcal{A}^{(n)}) \quad ;$$

let $\mu^{(n)}$ and $\nu^{(n)}$ be two probability measures on $(X^{(n)}, \mathcal{A}^{(n)})$, $\mu = \otimes \mu^{(n)}$, $\nu = \otimes \nu^{(n)}$; then μ and ν are disjoint iff the infinite product $\prod_{n=1}^{\infty} \int\sqrt{d\mu^{(n)} . d\nu^{(n)}} = 0$. This last result, plus the fact that μ and ν are either disjoint or equivalent, is due to Kakutani [23] . Moreover by the martingale theorem, if μ and ν are equivalent we have almost everywhere

$$\frac{d\mu}{d\nu} = \otimes \frac{d\mu^{(n)}}{d\nu^{(n)}} \quad .$$

Appendix I. DESINTEGRATION OF CONTINUOUS COCYCLES

Proposition I.1. Let G be a separable locally compact group, μ a bounded measure on a standard Borel space T , $K = \int^{\oplus} K_t . d\mu(t)$ a separable Hilbert integral, $A = \int^{\oplus} A_t . d\mu(t)$ a decomposable representation of G in K , b a continuous cocycle for A. There exists for every t a continuous cocycle b_t for A_t such that for every $g \in G$ we have $b(g) = \int^{\oplus} b_t(g) . d\mu(t)$.

We first prove several lemmas.

Lemma I.1. Let X be a locally compact space with a positive Radon measure ν , T, μ , K as in the above statement, u^1, u^2,... elements of K such that for every t the components u^i_t constitute an orthonormal basis of K_t ; let f be a continuous mapping of X into K. One can choose for every $x \in X$ a decomposition $f(x) = \int^{\oplus} f(x)_t . d\mu(t)$ such that the mapping $(x,t) \longmapsto (f(x)_t \mid u^i_t)$ is measurable for every i.

We can assume $T = [0,1[$; for every positive integer n and every $k = 0, 1, \ldots 2^n - 1$ we set $T^n_k = [k/2^n, (k+1)/2^n[$. We choose arbitrary decompositions $f(x) = \int^{\oplus} f'(x)_t . d\mu(t)$ and we define functions g^i_n on $X \times T$ by

$$g^i_n(x,t) = (\mu(T^n_k))^{-1} . \int_{T^n_k} (f'(x)_{t'} \mid u^i_{t'}) . d\mu(t')$$

$$= (\mu(T^n_k))^{-1} . (f(x) \mid P_{T^n_k} u^i)$$

for every $t \in T^n_k$; g^i_n is continuous with respect to x and stepwise with respect to t , hence it is measurable ; the set E of all pairs (x,t) such that the sequence $n \longmapsto g^i_n(x,t)$ is convergent for every i is measurable. By virtue of Lebesgue-Vitali theorem, for every x, $g^i_n(x,t)$ converges to $(f'(x)_t \mid u^i_t)$ $\forall$ i for almost every t, say for every t in some subset A_x with negligible complementary ; we have $E_x = \{ t \mid (x,t) \in E \} \supset A_x$ hence the complementary of E is negligible. For every $(x,t) \in E$, $g_n^i(x,t)$ converges to some measurable function $g^i(x,t)$ and we have $g^i(x,t) = (f'(x)_t \mid u_t^i)$ for $t \in A_x$. The function $\sum_i |g^i(x,t)|^2$ is measurable and the set F where it is finite is measurable ; for $t \in A_x$ we have

$$\sum_i | g^i(x,t)|^2 = \sum_i |(f'(x)_t \mid u_t^i)|^2 = \| f'(x)_t \|^2$$

whence $A_x \subset F_x$, the complementary of F_x is negligible. We set

$$f(x)_t = \begin{cases} \sum_i g^i(x,t).u_t^i & \text{for } t \in F_x \\ 0 & \text{for } t \notin F_x \text{ ;} \end{cases}$$

we have $f(x)_t = f'(x)_t$ for $t \in A_x$, whence $f(x) = \int^{\oplus} f(x)_t.d\mu(t)$; moreover the function

$$(x,t) \longmapsto (f(x)_t \mid u_t^i) = g^i(x,t) \qquad \text{for } (x,t) \in F$$

is measurable.

Lemma I.2. Let G, T, μ, K, A as in the statement of the proposition ; for every u and v in K the function $(g,t) \longmapsto (A_t(g).u_t \mid v_t)$ is measurable.

Note. Here and in the sequel we choose a left Haar measure ν on G.

Proof. The function $(A_t(g).u_t \mid v_t)$ is continuous with respect to g and measurable with respect to t ; then our lemma is a consequence of lemma 9.2 in G.W. Mackey, Induced representations of locally compact groups.I. Ann.Math.,t.55, 1952, p.101-139.

Lemma I.3. (i) Let G be a separable locally compact group, A a representation of G in a separable Hilbert space K, b a measurable mapping of G into K such that $b(gh) = b(g) + A(g).b(h)$ for almost all pairs (g,h). Then b is equal almost everywhere to a continuous cocycle.

(ii) Let (T,μ) be a standard measure space, $t \longmapsto A_t$ a measurable field of representations of G in K ; for every $t \in T$ consider a mapping $b_t : G \longrightarrow K$ with the same properties as b in (i) ; suppose that the mapping $(g,t) \longmapsto b_t(g)$ is measurable. Then the mapping $(g,t) \longmapsto b_t'(g)$ is measurable too, where b_t' is the continuous cocycle equal a.e. to b_t .

(The proof is inspired from [1], lemma 6.3.)

a) There exists a subset Γ in G with negligible complementary such that for every $g \in \Gamma$ we have

$$b(gh) = b(g) + A(g).b(h) \tag{1}$$

for all h in some subset U_g with negligible complementary.

Choose a nonnegligible integrable subset M in G and a compact neighbourhood N of e such that $\nu(M \cap \complement hM) \leq \nu(M)/2 \quad \forall\, h \in N$ (such an N does exist since the mapping $h \longmapsto \chi_{hM}$ is continuous from G in $L^1(G)$). We choose symmetric neighbourhoods N_1 and N_2 such that $N_1^2 \subset N$, $N_2^2 \subset N_1$.

b) We shall prove that b is bounded on $N \cap \Gamma$. For every real number a we set

$$E_a = \{ g \in G \mid \| b(g) \| < a \}$$

$$w(M,a) = \nu(M \cap E_a)/\nu(M) ;$$

we have

$$\lim_{a=+\infty} w(M,a) = 1 \tag{2}$$

since $\cup E_a = G$. For every $h \in \Gamma$ we have

$$w(hM,a) \leq 1 - w(M, \| b(h) \| - a) . \tag{3}$$

In fact for every $g \in U_h$ we have

$$b(hg) = b(h) + A(h).b(g)$$

$$\|b(g)\| = \|A(h).b(g)\| \geqslant \|b(h)\| - \|b(hg)\| ;$$

$g \in U_h$ and $\|b(hg)\| < a$ imply $\|b(g)\| > \|b(h)\| - a$; hence

$$\nu(hM \cap E_a) = \nu(\{ g \in M \cap U_h \mid \|b(hg)\| < a \})$$
$$\leq \nu(M \cap \complement E_{b(h) -a})$$

which implies (3).

For every $h \in N \cap \Gamma$ we have

$$w(M,a) = (\nu(M))^{-1}.[\nu(M \cap hM \cap E_a) + \nu(M \cap \complement hM \cap E_a)]$$
$$\leq w(hM,a) + \tfrac{1}{2}$$
$$\leq 3/2 - w(M, \|b(h)\| - a) .$$

Suppose b is not bounded on $N \cap \Gamma$; there exist $h_n \in N \cap \Gamma$ such that $\|b(h_n)\| \longrightarrow +\infty$; by (2) we have $w(M, \|b(h_n)\| - a) \longrightarrow 1$ whence $w(M,a) \leq \frac{1}{2}$ for every a, which contradicts (2).

c) We shall now prove that for every $g \in G$, b is essentially bounded on gN_1. First take g in Γ ; when h runs over $N \cap \Gamma \cap U_g$, gh runs over the set $gN \cap g\Gamma \cap gU_g$, the complementary of which in gN is negligible ; moreover (1) shows that $b(gh)$ is bounded ; hence b is essentially bounded on gN. Take now g in G ; since Γ is dense in G there exists an element g' in $\Gamma \cap gN_1$; then $g \in g'N_1$, $gN_1 \subset g'N_1^2 \subset g'N$ and b is essentially bounded on gN_1.

d) We are now in a position to prove part (i) of the lemma. Let us choose a function f on G which is continuous, $\nu(f) = 1$ and $\operatorname{supp} f \subset N_2$. The functions $h \longmapsto f(h).b(h)$ and $h \longmapsto f(gh).A(g).b(h)$ are integrable ; the function $g \longmapsto \int f(gh).A(g).b(h).dh$ is continuous [in fact suppose $g_n \longrightarrow g$ and $g_n \in N_2 g$; the supports of the functions $h \longmapsto f(g_n h).A(g_n).b(h)$ are included in the compact $g^{-1}N_1$ and these functions are uniformly bounded ; our assertion follows from Lebesgue convergence theorem]. Consider the continuous function

$$b'(g) = \int f(h).b(h).dh - \int f(gh).A(g).b(h).dh$$
$$= \int f(gh)\,[b(gh) - A(g).b(h)].dh\ ;$$

if g is in Γ, (1) shows that $b'(g) = b(g)$; thus b is equal a.e. to the continuous function b'. We have now to show that b' is a cocycle ; the set E of all pairs (g,h) such that $b'(gh) = b'(g) + A(g).b'(h)$ is closed, hence measurable ; for every $g \in \Gamma$ we have for almost all h

$$b'(gh) = b(gh) = b(g) + A(g).b(h) = b'(g) + A(g).b'(h)\ ;$$

hence for every $g \in \Gamma$, (g,h) belongs to E for almost every h ; $\complement E$ is negligible, hence empty, and b' is a cocycle.

e) Let us now prove part (ii) of the lemma. We can choose independantly of t the neighbourhood N and consequently N_1, N_2 and f ; we thus have

$$b'_t(g) = \int f(h).b_t(g).dh - \int f(gh).A_t(g).b_t(h).dh\ ;$$

our assertion follows from the fact that we can approach $b'_t(g)$ by linear combinations of functions of the form

$$\sum_i f(h_i).b_t(g) - \sum_i f(gh_i).A_t(g).b_t(h_i)$$

which are measurable by lemma 2 .

<u>Proof of proposition</u> I.1.

We can assume that the field $t \longmapsto K_t$ is a constant one. Let us choose u^i, u^i_t as in lemma 1 ; by this lemma there exist decompositions $b(g) = \int^{\oplus} b(g)_t . d\mu(t)$ such that the function $(g,t) \longmapsto (b(g)_t \mid u^i_t)$ is measurable for every i ; by lemma 2 the function $(g,t) \longmapsto (A_t(g).u^i_t \mid u^j_t)$ is measurable for every i and j ; then the function

$$(g,h,t) \longmapsto (A_t(g).b(h)_t \mid u_t^i) = (b(h)_t \mid A_t(g^{-1}).u_t^i)$$
$$= \sum_j (b(h)_t \mid u_t^j).\overline{(A_t(g^{-1}).u_t^i \mid u_t^j)}$$

is measurable too ; so is the function $(g,h,t) \longmapsto (b(gh)_t \mid u_t^i)$ and thence the function

$$(g,h,t) \longmapsto (b(gh)_t - b(g)_t - A_t(g).b(h)_t \mid u_t^i) ;$$

the set E of all triples (g,h,t) where this function is null for every i is measurable ; for every (g,h) , $\{ t \mid (g,h,t) \in E \}$ has a negligible complementary, which shows that E has a negligible complementary ; then for almost all t, say for $t \in T'$, we have

$$(b(gh)_t - b(g)_t - A_t(g).b(h)_t \mid u_t^i) = 0 \quad \forall\ i$$
$$b(gh)_t - b(g)_t - A_t(g).b(h)_t = 0 .$$

On the other hand for almost all t, say for $t \in T'' \subset T'$, the function $g \longmapsto b(g)_t$ is measurable ; for $t \in T''$ we can apply lemma 3 (i) : there exists a continuous cocycle b_t such that $b_t(g) = b(g)_t$ for almost all g ; for $t \notin T''$ we set $b_t = 0$. We must now prove that for every g we have $b_t(g) = b(g)_t$ for almost all t ; by lemma 3 (ii) the mapping $(g,t) \longmapsto b_t(g)$ is measurable ; for almost every t we have $b_t(g) = b(g)_t$ for almost all g ; hence $b_t(g) = b(g)_t$ for almost all pairs (g,t) ; hence for almost all g, say for $g \in F$, we have $b_t(g) = b(g)_t$ for almost all t. Take an arbitrary g in G ; g is the limit of a sequence $g_n \in F$; $b(g_n)$ converges to $b(g)$; there exists a subsequence (n_i) such that $b(g_{n_i})_t \longrightarrow b(g)_t$ a.e.; but $b(g_{n_i})_t = b_t(g_{n_i})$ a.e.; hence $b_t(g_{n_i}) \longrightarrow b(g)_t$ a.e.; but $b_t(g_{n_i}) \longrightarrow b_t(g)$ since b_t is continuous ; hence $b_t(g) = b(g)_t$ a.e.

Remark (due to K.Schmidt). The proof can be shortened : actually we do not need parts a),b),c) of lemma I.3 ; in fact we know by Fubini's theorem, that for any compact C in G and for almost every t the function $\| b(g)_t \|$ is integrable on C.

NOTATION INDEX

$\underline{N}$, $\underline{Z}$, $\underline{R}$, $\underline{C}$, $\underline{U}$ are respectively the set of positive integers, the set of positive or negative integers, the set of real numbers, the set of complex numbers , the set of complex numbers with modulus 1 ; $\underline{C}^*$ is the set of the non zero complex numbers.

TERMINOLOGICAL INDEX

BIBLIOGRAPHY

1 H.Araki. Factorizable Representation of Current Algebra. Publ. R.I.M.S. Kyoto Univ., t.5, 1970, p.361-422.

2 H.Araki-E.J.Woods. Complete Boolean Algebras of Type I factors. Publ.R.I.M.S. Kyoto Univ., t.2, 1966, p.157-242.

3 N.Bourbaki. Théorie des Ensembles.

4 N.Bourbaki. Intégration. ch.IX.

5 P.Cartier. Processus aléatoires généralisés. Séminaire Bourbaki n° 272, 1963/4.

6 J.L.Cathelineau. Sur les produits continus de nombres complexes. C.R.Acad.Sci., Paris, t.269, 1969, p.908-911.

7 J.M.Cook. The mathematics of second quantization. Trans.Amer. Math.Soc., t.74, 1953, p.222-245.

8 J.Dixmier. Les algèbres d'opérateurs dans l'espace hilbertien. Gauthier-Villars. 1957

9 J.Dixmier. Les C^* - algèbres et leurs représentations. Gauthier-Villars. 1964.

10 J.L.Doob. Stochastic Processes. J.Wiley. 1953.

11 J.Feldman. Decomposable Processes and Continuous Products of Probability Spaces. J.Funct.Anal.,t.8, 1971, p.1-51.

12 I.M.Gelfand-N.Ia.Vilenkin. Generalized Functions, t.4, Acad. Press, 1965.

13 A.Guichardet. Algèbres d'observables associées aux relations

de commutation. A.Colin. 1968.

14 A.Guichardet. Tensor Products of C^* - algebras. Mat.Institut, Aarhus Univ., Lecture Notes n° 12, 1969.

15 A.Guichardet. Sur la cohomologie des groupes topologiques. To appear in Bull.Sci.Math.

16 A.Guichardet. Special Topics in Topological Algebras. Gordon and Breach. 1968

17 A.Guichardet. Calcul Intégral. A.Colin. 1970.

18 A.Guichardet-A.Wulfsohn. Sur les produits tensoriels continus d'espaces hilbertiens. J.Funct.Anal., t.2, 1968, p.371-377.

19 P.R.Halmos. Lectures on Boolean Algebras. Van Nostrand. 1963.

20 K.Harzallah. Sur une démonstration de la formule de Lévy-Khinchin. Ann.Inst.Fourier. t.19, 1969, p.527-532.

21 E.Hewitt-L.J.Savage. Symmetric Measures on Cartesian Products. Trans.Amer,Math.Soc., t.80, 1955, p.470-501.

22 S.Johansen. An application of extreme points method to the representation of infinitely divisible distributions. Zeitsch. Wahrsch.Theorie, t.5, 1966, p.304-316.

23 S.Kakutani. On equivalence of infinite product measures. Ann. Math., t.49, 1948, p.214-224.

24 G.Kallianpur. The Role of Reproducing Kernel Hilbert Spaces in the Study of Gaussian Processes. Advances in Proba., t.2, M. Dekker. New York. 1970.

25 J.R.Klauder. Exponential Hilbert Space : Fock Space Revisited. J.Math.Phys., t.11-1, 1970, p.609-630.

26 G.W.Mackey. Unitary Representations of Group Extensions. Acta Math., t.99, 1958, p.265-311.

27 S.Mac Lane. Homology. Acad.Press. 1963.

28 P.A.Meyer. Probabilités et Potentiel. Paris. Hermann. 1966.

29 B.Mitchell. Theory of Categories. Acad.Press. 1965.

30 J.Neveu. Processus aléatoires gaussiens. Sém.de Math.Sup., Montréal. 1968.

31 K.R.Parthasarathy-R.R.Rao-S.R.S.Varadhan. Probability Distributions on Locally Compact Groups. Illinois J.Math., t.7,1963, p.337-369.

32 K.R.Parthasarathy. Probability Measures on Metric Spaces. Acad.Press. 1967.

33 K.R.Parthasarathy. Infinitely Divisible Representations and Positive Definite Functions on a Compact Group. Commun.Math. Phys., t.16, 1970, p.148-156.

34 K.R.Parthasarathy-K.Schmidt. Infinitely Divisible Projective Representations, Cocycles and Levy-Khinchin-Araki Formula on Locally Compact Groups. To appear.

35 K.R.Parthasarathy-K.Schmidt. Factorizable Representations of Current Groups and the Araki-Woods Imbedding Theorem. To appear.

36 F.Rocca. Cohérence optique et convolution gauche, un théorème d'équivalence optique. J.Phys., t.28, 1967, p.113-119.

37 V.A.Rohlin. Selected Problems in the Metric Theory of Dynamical Systems (russian). Uspekhi Mat.Nauk, t.4, 1949, p.57-128.

38 V.V.Sazonov-V.N.Tutubalin. Probability Distributions on Topological Groups. Th.of Proba.and Appl., 1966, t.11, p.1-45.

39 I.E.Segal. Tensor Algebras over Hilbert spaces.I. Trans.Amer. Math.Soc., t.81, 1956, p.106-134.

40 I.E.Segal. Ergodic Subgroups of the Orthogonal Group on a Real Hilbert space. Ann.Math., t.66, 1957, p.297-303.

41 R.Sikorski. Boolean Algebras, Erg. der Math., vol.25, Springer 1964.

42 H.Totoki. Ergodic Theory. Mat.Institut, Aarhus Univ., Lecture Notes n° I4, 1969.

43 Y.Umemura. Measures on Infinite Dimensional Vector Spaces. Publ.R.I.M.S.Kyoto Univ., t.1, 1965, p.1-47.

44 V.Varadarajan. Geometry of Quantum Theory. Van Nostrand, 1970.

Please turn over

Vol. 146: A. B. Altman and S. Kleiman, Introduction to Grothendieck Duality Theory. II, 192 pages. 1970. DM 18,–

Vol. 147: D. E. Dobbs, Cech Cohomological Dimensions for Commutative Rings. VI, 176 pages. 1970. DM 16,–

Vol. 148: R. Azencott, Espaces de Poisson des Groupes Localement Compacts. IX, 141 pages. 1970. DM 16,–

Vol. 149: R. G. Swan and E. G. Evans, K-Theory of Finite Groups and Orders. IV, 237 pages. 1970. DM 20,–

Vol. 150: Heyer, Dualität lokalkompakter Gruppen. XIII, 372 Seiten. 1970. DM 20,–

Vol. 151: M. Demazure et A. Grothendieck, Schémas en Groupes I. (SGA 3). XV, 562 pages. 1970. DM 24,–

Vol. 152: M. Demazure et A. Grothendieck, Schémas en Groupes II. (SGA 3). IX, 654 pages. 1970. DM 24,–

Vol. 153: M. Demazure et A. Grothendieck, Schémas en Groupes III. (SGA 3). VIII, 529 pages. 1970. DM 24,–

Vol. 154: A. Lascoux et M. Berger, Variétés Kähleriennes Compactes. VII, 83 pages. 1970. DM 16,–

Vol. 155: Several Complex Variables I, Maryland 1970. Edited by J. Horváth. IV, 214 pages. 1970. DM 18,–

Vol. 156: R. Hartshorne, Ample Subvarieties of Algebraic Varieties. XIV, 256 pages. 1970. DM 20,–

Vol. 157: T. tom Dieck, K. H. Kamps und D. Puppe, Homotopietheorie. VI, 265 Seiten. 1970. DM 20,–

Vol. 158: T. G. Ostrom, Finite Translation Planes. IV. 112 pages. 1970. DM 16,–

Vol. 159: R. Ansorge und R. Hass. Konvergenz von Differenzenverfahren für lineare und nichtlineare Anfangswertaufgaben. VIII, 145 Seiten. 1970. DM 16,–

Vol. 160: L. Sucheston, Constributions to Ergodic Theory and Probability. VII, 277 pages. 1970. DM 20,–

Vol. 161: J. Stasheff, H-Spaces from a Homotopy Point of View. VI, 95 pages. 1970. DM 16,–

Vol. 162: Harish-Chandra and van Dijk, Harmonic Analysis on Reductive p-adic Groups. IV, 125 pages. 1970. DM 16,–

Vol. 163: P. Deligne, Equations Différentielles à Points Singuliers Reguliers. III, 133 pages. 1970. DM 16,–

Vol. 164: J. P. Ferrier, Seminaire sur les Algebres Complètes. II, 69 pages. 1970. DM 16,–

Vol. 165: J. M. Cohen, Stable Homotopy. V, 194 pages. 1970. DM 16,–

Vol. 166: A. J. Silberger, PGL_2 over the p-adics: its Representations, Spherical Functions, and Fourier Analysis. VII, 202 pages. 1970. DM 18,–

Vol. 167: Lavrentiev, Romanov and Vasiliev, Multidimensional Inverse Problems for Differential Equations. V, 59 pages. 1970. DM 16,–

Vol. 168: F. P. Peterson, The Steenrod Algebra and its Applications: A conference to Celebrate N. E. Steenrod's Sixtieth Birthday. VII, 317 pages. 1970. DM 22,–

Vol. 169: M. Raynaud, Anneaux Locaux Henséliens. V, 129 pages. 1970. DM 16,–

Vol. 170: Lectures in Modern Analysis and Applications III. Edited by C. T. Taam. VI, 213 pages. 1970. DM 18,–

Vol. 171: Set-Valued Mappings, Selections and Topological Properties of 2^X. Edited by W. M. Fleischman. X, 110 pages. 1970. DM 16,–

Vol. 172: Y.-T. Siu and G. Trautmann, Gap-Sheaves and Extension of Coherent Analytic Subsheaves. V, 172 pages. 1971. DM 16,–

Vol. 173: J. N. Mordeson and B. Vinograde, Structure of Arbitrary Purely Inseparable Extension Fields. IV, 138 pages. 1970. DM 16,–

Vol. 174: B. Iversen, Linear Determinants with Applications to the Picard Scheme of a Family of Algebraic Curves. VI, 69 pages. 1970. DM 16,–

Vol. 175: M. Brelot, On Topologies and Boundaries in Potential Theory. VI, 176 pages. 1971. DM 18,–

Vol. 176: H. Popp, Fundamentalgruppen algebraischer Mannigfaltigkeiten. IV, 154 Seiten. 1970. DM 16,–

Vol. 177: J. Lambek, Torsion Theories, Additive Semantics and Rings of Quotients. VI, 94 pages. 1971. DM 16,–

Vol. 178: Th. Bröcker und T. tom Dieck, Kobordismentheorie. XVI, 191 Seiten. 1970. DM 18,–

Vol. 179: Seminaire Bourbaki – vol. 1968/69. Exposés 347-363. IV. 295 pages. 1971. DM 22,–

Vol. 180: Séminaire Bourbaki – vol. 1969/70. Exposés 364-381. IV 310 pages. 1971. DM 22,–

Vol. 181: F. DeMeyer and E. Ingraham, Separable Algebras over Commutative Rings. V, 157 pages. 1971. DM 16.–

Vol. 182: L. D. Baumert. Cyclic Difference Sets. VI, 166 pages. 1971 DM 16,–

Vol. 183: Analytic Theory of Differential Equations. Edited by P. F. Hsieh and A. W. J. Stoddart. VI, 225 pages. 1971. DM 20,–

Vol. 184: Symposium on Several Complex Variables, Park City, Utah 1970. Edited by R. M. Brooks. V, 234 pages. 1971. DM 20,–

Vol. 185: Several Complex Variables II, Maryland 1970. Edited by J. Horváth. III, 287 pages. 1971. DM 24,–

Vol. 186: Recent Trends in Graph Theory. Edited by M. Capobianco, J. B. Frechen/M. Krolik. VI, 219 pages. 1971. DM 18.–

Vol. 187: H. S. Shapiro, Topics in Approximation Theory. VIII, 275 pages 1971. DM 22,–

Vol. 188: Symposium on Semantics of Algorithmic Languages. Edited by E. Engeler. VI, 372 pages. 1971. DM 26,–

Vol. 189: A. Weil, Dirichlet Series and Automorphic Forms. V, 16[illegible] pages. 1971. DM 16,–

Vol. 190: Martingales. A Report on a Meeting at Oberwolfach, May 17-23, 1970. Edited by H. Dinges. V, 75 pages. 1971. DM 16,–

Vol. 191: Séminaire de Probabilités V. Edited by P. A. Meyer. IV, 37[illegible] pages. 1971. DM 26,–

Vol. 192: Proceedings of Liverpool Singularities – Symposium I. Edited by C. T. C. Wall. V, 319 pages. 1971. DM 24,–

Vol. 193: Symposium on the Theory of Numerical Analysis. Edited by J. Ll. Morris. VI, 152 pages. 1971. DM 16,–

Vol. 194: M. Berger, P. Gauduchon et E. Mazet. Le Spectre d'une Variété Riemannienne. VII, 251 pages. 1971. DM 22,–

Vol. 195: Reports of the Midwest Category Seminar V. Edited by J.W. Gray and S. Mac Lane. III, 255 pages. 1971. DM 22,–

Vol. 196: H-spaces – Neuchâtel (Suisse)- Août 1970. Edited by F. Sigrist, V, 156 pages. 1971. DM 16,–

Vol. 197: Manifolds – Amsterdam 1970. Edited by N. H. Kuiper. V, 23[illegible] pages. 1971. DM 20,–

Vol. 198: M. Hervé, Analytic and Plurisubharmonic Functions in Finite and Infinite Dimensional Spaces. VI, 90 pages. 1971. DM 16.–

Vol. 199: Ch. J. Mozzochi, On the Pointwise Convergence of Fourier Series. VII, 87 pages. 1971. DM 16,–

Vol. 200: U. Neri, Singular Integrals. VII, 272 pages. 1971. DM 22,–

Vol. 201: J. H. van Lint, Coding Theory. VII, 136 pages. 1971. DM 16,–

Vol. 202: J. Benedetto, Harmonic Analysis on Totally Disconnected Sets. VIII, 261 pages. 1971. DM 22,–

Vol. 203: D. Knutson, Algebraic Spaces. VI, 261 pages. 1971 DM 22,–

Vol. 204: A. Zygmund, Intégrales Singulières. IV, 53 pages. 1971 DM 16,–

Vol. 205: Séminaire Pierre Lelong (Analyse) Année 1970. VI, 243 pages. 1971. DM 20,–

Vol. 206: Symposium on Differential Equations and Dynamical Systems. Edited by D. Chillingworth. XI, 173 pages. 1971. DM 16,–

Vol. 207: L. Bernstein, The Jacobi-Perron Algorithm – Its Theory and Application. IV, 161 pages. 1971. DM 16,–

Vol. 208: A. Grothendieck and J. P. Murre, The Tame Fundamental Group of a Formal Neighbourhood of a Divisor with Normal Crossings on a Scheme. VIII, 133 pages. 1971. DM 16,–

Vol. 209: Proceedings of Liverpool Singularities Symposium II. Edited by C. T. C. Wall. V, 280 pages. 1971. DM 22,–

Vol. 210: M. Eichler, Projective Varieties and Modular Forms. III, 11[illegible] pages. 1971. DM 16,–

Vol. 211: Théorie des Matroïdes. Edité par C. P. Bruter. III, 10[illegible] pages. 1971. DM 16,–

Vol. 212: B. Scarpellini, Proof Theory and Intuitionistic Systems. VII, 291 pages. 1971. DM 24,–

Vol. 213: H. Hogbe-Nlend, Théorie des Bornologies et Applications. V, 168 pages. 1971. DM 18,–

Vol. 214: M. Smorodinsky, Ergodic Theory, Entropy. V, 64 pages 1971. DM 16,–